Exercises for the Microbiology Laboratory

3rd EDITION

Burton E. Pierce
San Diego City College

Michael J. Leboffe
San Diego City College

Morton Publishing Company
925 W. Kenyon, Unit 12
Englewood, Colorado 80110
http://www.morton-pub.com

Book Team

Publisher	Douglas N. Morton
Biology Editor	David Ferguson
Cover & Design	Bob Schram, Bookends, Inc.
Composition	Ash Street Typecrafters, Inc.

Printed in the United States of America

10 9 8 7 6 5 4 3 2

ISBN: 0-89582-657-7

Preface

This edition of *Exercises for the Microbiology Manual* retains many of the features that have made previous editions successful. That is, it is a no-frills approach to introductory microbiology with exercises that work. Other successful features, such as recipes for each medium and references for each exercise remain.

The 3rd Edition continues to be supported, for those who want it, by the *Photographic Atlas for the Microbiology Laboratory, 3rd Ed.,* also available from Morton Publishing. As before, both works have a loose-leaf design, allowing integration of the photographs with the procedures.

This edition differs in several ways from its predecessors. There has been moderate reorganization and addition of more than a dozen new exercises (from the lab manual, *Microbiology Laboratory Theory and Application,* also available from Morton Publishing). All parts have been rewritten and clarified, where necessary, and the suggested organisms have been reevaluated and replaced with the goal of clearer results and reduction of organism inventory.

Following is a listing of major changes from the second edition.

Section 1 Fundamental Skills for the Microbiology Laboratory

- The long and tedious Exercise 1-2 (Aseptic Transfers) has been split (without reduction of content) into two exercises and three appendices. Commonly used methods remain in Section 1, whereas the more specialized ones have been moved to Appendices B, C, and D.
- We moved the streak plate method of isolation from Section 4 to Section 1.

Section 2 Microbial Growth

- Nine new exercises have been added to this section, mostly in the areas of environmental factors affecting growth and microbial control.
- The Kirby-Bauer test was moved to this section.

Section 3 Microscopy and Staining

- The exercise on introductory microscopy has been expanded.
- The morphological unknown has been moved from its former section into this one.

SECTION 4 SELECTIVE MEDIA

- An exercise on Endo Agar was added.

SECTION 5 DIFFERENTIAL TESTS

- Exercises for Malonate broth and β-lactamase test have been added.
- The exercise on Kligler's Iron Agar has been removed.
- The biochemical unknown was moved into this section.

SECTION 7 ENVIRONMENTAL, FOOD, AND MEDICAL MICROBIOLOGY

- This section no longer contains exercises for molecular biology. These were moved to Section 8.
- Seven new exercises have been added.

SECTION 8 MICROBIAL GENETICS

- This section now includes the exercises (Ames test and UV Damage and Repair) related to mutation.
- A new exercise on DNA extraction has been added.

APPENDICES

- New appendices on use of the spectrophotometer and digital pipettors have been added.

It is our intention to provide you with a set of useful and interesting exercises that will make your microbiology experience a positive one. We appreciate and encourage your comments.

Burton E. Pierce
Michael J. Leboffe

Acknowledgments

To all of you, past and present, who helped make this manual possible we offer you our heartfelt gratitude. Let this page recognize and make public your kindness and unselfish contribution to our success.

From San Diego City College: Thanks to Terence Burgess, President, Ron Manzoni, Vice President of Instruction, Dr. Marianne Tortorici, Dean of Science, Nursing, Health and Athletics, and Carol Dexheimer, Business Manager who made campus facilities available to us through the Civic Center Program. Thanks also to Dr. Dianne Anderson, Susan Antoniades, James Bartley, David Brady, Dr. Emily Burke, Dr. Joyce Costello, Paul Detwiler, Julie Haugsness-White, Dr. Anita Hettena, James Kapin, Dr. Khasha Komejany, Dr. Jean Nichols, Debra Reed, Erin Rempala-Kim, Brett Ruston, Dr. David Singer, Dr. Minou Djawdan Spradley, and Gary Wisehart from our own Biology Department family for enduring the inconvenience and clutter created over the past year.

We again acknowledge Dr. David Singer and James Bartley from the San Diego City College Biology Department for constructing and developing earlier versions of many of the exercises that appear herein. We know we stand on your shoulders. Thanks to the San Diego City College Fall and Spring 2003-2004 day and evening Microbiology classes for helping us develop several exercises.

Thanks to Ellen Meyer from the American Public Health Association for permission to use table 9222:I from *Standard Methods for the Examination of Water and Wastewater, 20th Ed.*

We also acknowledge the helpful suggestions of the following reviewers of previous work: Dr. Donald Gerbig Jr., Kent State University-Tuscarawas, Rob Hubert, Iowa State University, Bobbie Pettriess at Wichita State University, and Dr. Kate Richardson, Portland Community College–Sylvania Campus.

Thanks again to Doug Morton, Chrissy Morton DeMier, David Ferguson and the staff of Morton Publishing Company for your constant support and encouragement. We are greatly indebted to Bob Schram from Bookends Publications and Joanne Saliger, Elaine McFarlane, and Patricia Govro from Ash Street Typecrafters, Inc. for their attractive design and layout.

Thanks finally to our families, Michele Pierce, and Karen, Nathan, Alicia, and Eric Leboffe for their patience and support during the past year.

Contents

Introduction to Safety and Laboratory Guidelines

Microbiology lab can be an interesting and exciting experience, but there are also potential hazards of which you should be aware. Improper handling of chemicals, equipment, and/or microbial cultures is a dangerous practice and can result in injury or infection. Safety with lab equipment will be covered when you first use that particular piece of equipment, as will specific examples of chemical safety. Our main concern at this point is introducing you to safe handling and disposal of microbes.*

Microorganisms present varying degrees of risk to laboratory personnel (students, technicians, and faculty), people outside the laboratory, and the environment, so it is imperative that microbial cultures be handled safely.

Please follow these and any other safety guidelines required by your college.

Student Conduct

1. To reduce the risk of infection, do not smoke, eat, drink, or bring food or drinks into the laboratory room—even if lab work is not being done.
2. Application of cosmetics and insertion or removal of contact lenses in the laboratory are not permitted.
3. Wash your hands *thoroughly* with soap and water after handling living microbes and before leaving the laboratory each day. Also, wash hands after removing gloves.
4. Lab time is precious so come to lab prepared for that day's work. Figuring out what to do as you go is an undertaking likely to produce confusion and accidents.
5. Do not remove any organisms or chemicals from the laboratory.
6. Work carefully and methodically. Do not hurry through any laboratory procedure.

Basic Laboratory Safety

1. Wear protective clothing (*i.e.,* a lab coat) in the laboratory when handling microbes. It should be removed prior to leaving the lab and autoclaved regularly.
2. Wear eye protection whenever heating chemicals. Eye protection is also recommended for those wearing contact lenses.

---

* Your instructor may augment or revise these guidelines to fit the conventions of your laboratory.

3. Turn off your Bunsen burner when not in use. Not only is it a fire and safety hazard, but an unnecessary source of heat in the room.
4. Tie back long hair. It is a potential source of contamination as well as a likely target for fire.
5. If you are feeling ill (for whatever reason) do not work with live microbes. There are other ways you may contribute to your lab group (*e.g.*, record data, fill out culture labels, retrieve equipment, *etc.*).
6. If you are pregnant or are taking immunosuppressant drugs, please see the instructor. It may be in your best *long*-term interests to postpone taking this class.
7. Wear disposable gloves while staining microbes and handling blood products—plasma, serum, antiserum, or whole blood. Handling blood can be hazardous even with gloves. Consult your instructor before attempting to work with any blood products.
8. Use an antiseptic (*e.g.*, Betadine) on your skin if it is exposed to a spill containing microorganisms. Your instructor will tell you which antiseptic you will be using.
9. Never pipette by mouth. Always use mechanical pipettors (Figure D-1 in Appendix D).
10. Dispose of uncontaminated broken glass in a "sharps" container.
11. Use a fume hood to perform any work involving highly volatile chemicals or stains that need to be heated.
12. Find the first aid kit and make a mental note of its location.
13. Find the fire blanket and fire extinguisher, note their location and make a plan for how to get them in an emergency.
14. Find the eye wash basin, learn how to operate it, and remember its location.

Reducing Contamination of Self, Others, Cultures, and the Environment

1. Wipe the desk top with a disinfectant (*e.g.*, Amphyl or 10% chlorine bleach) before *and* after each lab period in which live organisms are used. Appropriate disinfectant will be supplied.
2. Never lay culture tubes down on the table; they should always remain upright in a tube holder. Even solid media tubes contain moisture or condensation that may leak out and contaminate the work surface, your hands, or other cultures.
3. Cover any culture spills with paper towels. Immediately soak the towels with disinfectant and allow them to stand for 20 minutes. Report the spill to your instructor. When finished, place the towels in the container designated for autoclaving.
4. Place all nonessential books and papers under the desk. A cluttered lab table is an invitation for an accident, an accident that may contaminate your expensive school supplies. (We recommend bringing only relevant pages from this manual to each lab session.)
5. Place a disinfectant-soaked towel on the work area when pipetting microbial cultures. This reduces contamination and possible aerosols if a drop escapes from the pipette and hits the tabletop.

Disposal of Contaminated Materials

In most instances, the preferred method of sterilizing contaminated waste and reusable equipment is the autoclave.

1. Dispose of plate cultures (if plastic Petri dishes are used) and other contaminated nonreusable items in the appropriate container to be sterilized when you are finished with them.
2. Remove all labels from tube cultures and other contaminated reusable items and place them in the container designated for autoclaving.
3. Dispose of all blood product samples as well as disposable gloves in the container designated for autoclaving.
4. Place used microscope slides of bacteria in a container designated for autoclaving or soak them in disinfectant solution for at least 30 minutes before cleaning or discarding them. Follow your laboratory guidelines for glass disposal.
5. Place contaminated broken glass in the container designated for autoclaving. After sterilization, dispose of the glass in a "sharps" container or specialized broken glass container.

References

Barkley, W. Emmett and John H. Richardson. 1994. Chapter 29 in *Methods for General and Molecular Bacteriology,* edited by Philipp Gerhardt, R. G. E. Murray, Willis A. Wood, and Noel R. Krieg. American Society for Microbiology, Washington, DC.

Collins, C. H., Patricia M. Lyne and J. M. Grange. 1995. Chapters 1 and 4 in *Collins and Lyne's Microbiological Methods*, 7th Ed. Butterworth-Heineman, Oxford.

Darlow, H. M. 1969. Chapter VI in *Methods in Microbiology,* Vol. 1, edited by J. R. Norris and D. W. Ribbins. Academic Press, Ltd., London.

Fleming, Diane O. and Debra L. Hunt (Editors). 2000. *Laboratory Safety—Principles and Practices*, 3rd Ed. American Society for Microbiology, Washington, DC.

Koneman, Elmer W., Stephen D. Allen, William M. Janda, Paul C. Schreckenberger, and Washington C. Winn, Jr. 1997. *Color Atlas and Textbook of Diagnostic Microbiology*, 5th Ed. Lippincott-Raven Publishers, Philadelphia and New York.

Power, David A. and Peggy J. McCuen. 1988. Pages 2 and 3 in *Manual of BBL® Products and Laboratory Procedures*, 6th Ed. Becton Dickinson Microbiology Systems, Cockeysville, MD.

Richmond, Jonathan Y., and Robert W. McKinney (ed.). May 1999. U. S. Department of Health and Human Services, *Biosafety in Microbiological and Biomedical Laboratories*, 4th Ed. U. S. Government Printing Office, Washington, DC.

SECTION ONE

1 Fundamental Skills for the Microbiology Laboratory

Bacterial and fungal **cultures** are grown and maintained on or in solid and liquid substances called **media**. Preparation of these media involves weighing ingredients, measuring liquid volumes, calculating proportions, handling basic laboratory glassware, and operating a pH meter and an autoclave. In Exercise 1-1, you will learn and practice these fundamental skills by preparing a couple of simple growth media. When you have completed the exercises you will have the skills necessary to prepare most any medium if given the recipe.

A second fundamental skill necessary for any microbiologist is the ability to transfer bacterial cells from one place to another without contaminating the original culture, the new medium, or the environment (including the microbiologist). This **aseptic** (sterile) transfer technique is required for virtually all procedures in which living microbes are handled, including isolations, staining, and differential testing. Exercises 1-2 through 1-4 are devoted to descriptions of common transfer and inoculation methods. Less frequently used methods are covered in Appendices B through D.

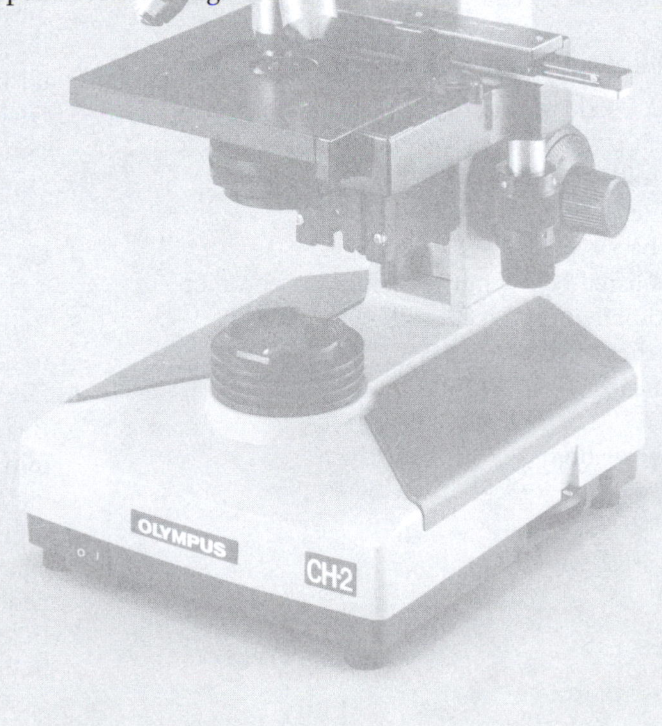

1-1 NUTRIENT AGAR AND NUTRIENT BROTH PREPARATION

Microbiologists use a variety of growth media to cultivate microbes. These media may be formulated from scratch, but are more typically produced by rehydrating commercially available powdered media. Media routinely encountered in the microbiology laboratory range from the widely used, general-purpose growth media, to the more specific selective and differential media used in identification of microbes. This exercise will teach you how to prepare simple general growth media.

Nutrient agar and nutrient broth are common media used for maintaining bacterial cultures. To be of practical use, they need to meet the diverse nutrient requirements of a majority of bacteria. As such, they are formulated from sources that supply carbon and nitrogen in a variety of forms—amino acids, purines, pyrimidines, mono- to polysaccharides, and various lipids. Generally, these are provided in digests of plant material (phytone) or animal material (peptone and others). Since the exact composition and amounts of carbon and nitrogen in these ingredients are unknown, general growth media are considered to be **undefined**.

While in most classes (due to limited time), media are prepared by a laboratory technician, it is instructive for novice microbiologists to at least gain exposure to what is involved in media preparation. Your instructor will provide specific instructions on how to execute this exercise using your particular laboratory equipment.

MATERIALS NEEDED FOR THIS EXERCISE

- Two or three 2 L Erlenmeyer flasks *per student group*
- Three or four 500 mL Erlenmeyer flasks and covers (can be aluminum foil)
- Stirring hotplate
- Magnetic stir bars
- All ingredients listed below in the recipes (or commercially prepared dehydrated media)
- Sterile Petri dishes
- Test tubes and caps
- Balance
- Weighing paper or boats
- Spatulas

RECIPES

Nutrient Agar

Beef extract	3.0 g
Peptone	5.0 g
Agar	15.0 g
Distilled or deionized water	1.0 L

pH 6.6–7.0 at 25°C

Nutrient Broth

Beef extract	3.0 g
Peptone	5.0 g
Distilled or deionized water	1.0 L

pH 6.6–7.0 at 25°C

MEDIA PREPARATION

Day One

To minimize contamination while preparing media, clean the work surface, turn off all fans, and close any doors that might allow excessive air movements.

Nutrient Agar Tubes

1. Weigh the ingredients on a balance.
2. Suspend the ingredients in one liter of distilled or deionized water in the two-liter flask, mix well, and boil until fully dissolved.
3. Dispense 7 mL portions into test tubes and cap loosely. If preparing agar deep stabs, use 10 mL per tube.
4. Sterilize the medium by autoclaving for 15 minutes at 121°C.
5. After autoclaving, cool to room temperature with the tubes in an upright position for agar deep tubes. Cool with the tubes on an angle for agar slants.
6. Incubate the slants or deep stabs at 37°C for 24 to 48 hours.

Nutrient Agar Plates

1. Weigh the ingredients on a balance.
2. Suspend the ingredients in one liter of distilled or deionized water in the two-liter flask, mix well, and boil until fully dissolved.
3. Divide into three or four 500 mL flasks for pouring. (Smaller flasks are easier to handle when pouring plates. Don't forget to add a magnetic stir bar to each flask before autoclaving.)
4. Cover the containers and autoclave for 15 minutes at 121°C to sterilize the medium.
5. Remove the sterile agar from the autoclave and allow it to cool to 50°C while stirring on a hotplate.
6. Dispense approximately 20 mL into sterile Petri plates **Be careful! The flask will still be hot so wear an oven mitt.** While you pour the agar, shield the Petri dish with its lid to reduce the chance of introducing airborne contaminants. If necessary, gently swirl each plate so the agar completely covers the bottom. Allow the agar to cool and solidify before moving the plates.
7. Invert the plates and incubate them at 37°C for 24 to 48 hours. (These plates are being used to check your aseptic technique while preparing media. If they were going to be used for inoculation, they should be inverted and stored on a counter top or in the incubator for 24 hours to allow them to dry prior to use.)

Nutrient Broth

1. Weigh the ingredients on a balance.
2. Suspend the ingredients in one liter of distilled or deionized water in the two-liter flask. Agitate and heat slightly (if necessary) to dissolve them completely.
3. Dispense 7.0 mL portions into test tubes and cap loosely.
4. Sterilize the medium by autoclaving for 15 minutes at 121°C.
5. Incubate the tubes at 37°C for 24 to 48 hours.

Day Two

1. Remove all plates and tubes from the incubator and record the number of each medium type you prepared.
2. Record the number of apparently sterile plates and tubes in table below.
3. Calculate your percentage of successful preparations for each. How was your technique?

REFERENCES

DIFCO Laboratories. 1984. Pages 619 and 622 in *DIFCO Manual*, 10th Ed. DIFCO Laboratories, Detroit, MI.

Power, David A. and Peggy J. McCuen. 1988. Pages 214 and 215 in *Manual of BBL® Products and Laboratory Procedures*, 6th Ed. Becton Dickinson Microbiology Systems, Cockeysville, MD.

■ OBSERVATIONS AND INTERPRETATIONS

Record your data in the table below. Speculate on probable sources of contamination, if any is observed.

OBSERVATIONS AND INTERPRETATIONS

MEDIUM	NUMBER OF STERILE PREPARATIONS	TOTAL NUMBER PREPARED	PERCENTAGE OF SUCCESSFUL PREPARATIONS	PROBABLE SOURCE(S) OF CONTAMINATION (IF ANY)
Nutrient Agar Tubes (Slant or Deep)				
Nutrient Agar Plates				
Nutrient Broths				

1-2 COMMON ASEPTIC TRANSFERS AND INOCULATION METHODS

As a microbiology student, you will be required to transfer living microbes from one place to another **aseptically** (*i.e.,* without contamination of the culture, the sterile medium, or the surroundings). While you won't be expected to master all transfer methods right now, you will be expected to perform most of them over the course of the semester. Refer back to this section as needed.

In order to prevent contamination of the sample, inoculating instruments (Figure 1-1) must be sterilized prior to use. Inoculating loops and needles are sterilized immediately before use in an incinerator or Bunsen burner flame. The mouths of tubes or flasks containing cultures or media are also incinerated at the time of transfer. Instruments that are not conveniently or safely incinerated, such as Pasteur pipettes, cotton applicators, glass pipettes, and digital pipettor tips, are sterilized inside wrappers or containers by autoclaving prior to use.

FIGURE 1-1 Inoculating Tools
Any of several different instruments may be used to transfer a microbial sample, the choice of which depends on the sample source, its destination, and any special requirements imposed by the specific protocol. Shown here are several examples of transfer instruments. From left to right: serological pipette (see Appendix C), disposable transfer pipette, Pasteur pipette, inoculating needle, inoculating loop, disposable inoculating needle/loop, cotton swab (see Appendix B and Exercise 1-3), and glass spreading rod (see Exercise 1-4).

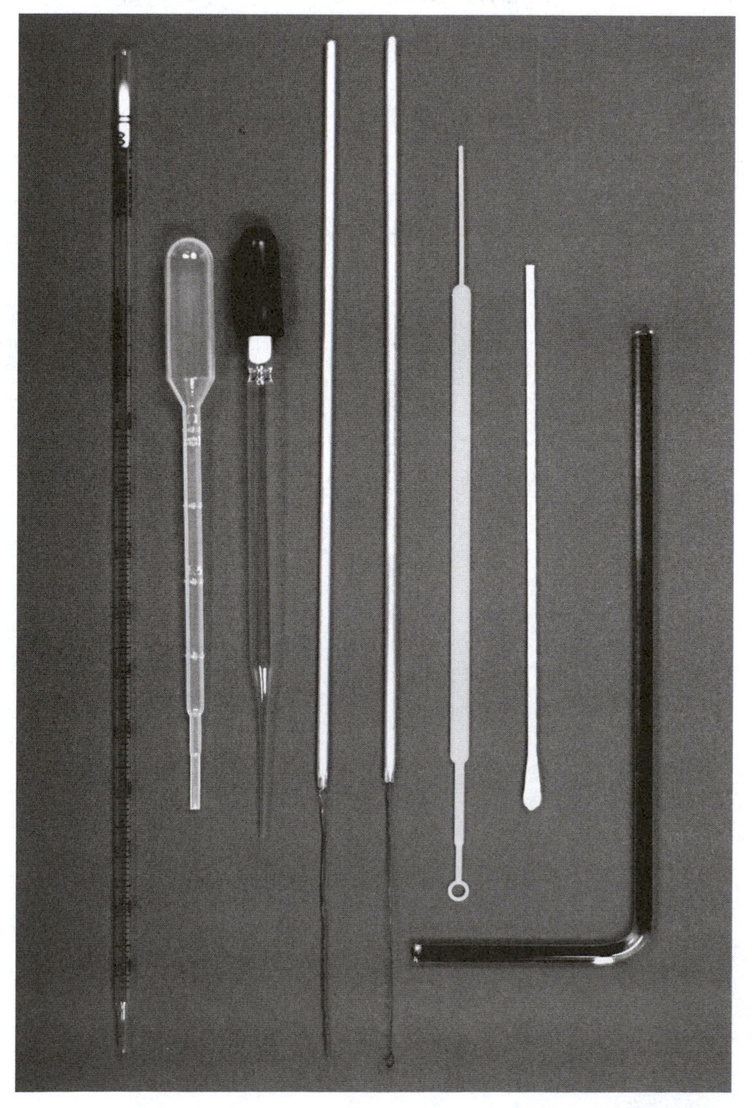

Aseptic transfers are not difficult; however, a little preparation will help assure a safe and successful procedure. Before you begin, you will need to know where the sample is coming from, its destination, and the type of transfer instrument to be used. This exercise provides a step-by-step description of many different transfer methods. Some skills are basic to most transfers and are described in detail once (in "The Basics" section), but, in an effort to avoid too much repetition, are only briefly mentioned in the "Specific Transfer Methods" section. These are printed in regular type. Other skills are more specialized and are introduced as part of the particular transfer. These are printed in bold type. Certain less routine transfer methods are discussed in Appendices B through D.

THE BASICS

This is a listing of general techniques and practices.

1. Be organized. Arrange all media in advance and clearly label them. Be sure not to place the labels in such a way as to obscure your view of the inside of the tube or plate.
2. Take your time. Work efficiently, but *do not hurry*. You are handling potentially dangerous microbes.
3. All media tubes should be in a test tube rack when not in use whether or not they are sterile. Tubes should never be laid on the table surface (Figure 1-2).
4. Hold the handle of an inoculating needle or loop like a pencil in your dominant hand and relax! (Figure 1-2)
5. Adjust your Bunsen burner so its flame has an inner and an outer cone.
6. Sterilize a loop/needle by incinerating it in the Bunsen burner flame (Figure 1-3). Pass it through the tip of the flame's inner cone, holding it at an angle with the loop end pointing downward. Begin flaming about 2 cm up the handle, then proceed down the wire by pulling the loop backward through the flame until the entire wire has become uniformly red-hot. Flaming in this direction limits aerosol production by allowing the tip to heat up more slowly than if it were thrust into the flame immediately.
7. Hold a culture tube in your nondominant hand and move it, not the loop, as you transfer. This will minimize aerosol production from loop movement.
8. Grasp the tube's cap with your little finger and remove it by pulling the *tube* away from the cap. Hold the cap in your little finger during the transfer (Figure 1-4).

(The cap should be loosened prior to transfer, especially if it's a screw top cap.) When replacing the cap, move the tube back to the cap in order to keep your loop hand still. The replaced cap doesn't need to be on firmly yet—just enough to cover the tube.

9. Hold open tubes on an angle to minimize the chance of airborne contamination.
10. Flame tubes by passing the open end through the Bunsen burner flame two or three times (Figure 1-5).
11. Suspend bacteria in a broth with a vortex mixer prior to transfer (Figure 1-6). Be sure not to mix so vigorously that broth gets into the cap or that you lose control of the tube. It's best to start slowly, then gently increase the speed until the tip of the vortex reaches the bottom of the tube. Alternatively, broth may be agitated by drumming your fingers along the length of the tube several times (Figure 1-7). Be careful not to splash the broth into the cap or lose control of the tube.
12. When opening a plate, use the lid as a shield to minimize the chances of airborne contamination. (Figure 1-8).
13. Label all culture tubes and plates with your name, the date, the medium, and the inoculum.

Specific Transfer Methods

There are two basic stages in transfers: 1) obtaining the sample to be transferred, and 2) transferring to the sterile culture medium. These may be combined in various ways. The following descriptions are organized to reflect that flexibility.

■ FIGURE 1-2 Microbiologist at Work
Materials are neatly positioned and not in the way. To prevent spills, culture tubes are stored upright in a test tube rack—they are never laid on the table. The microbiologist is relaxed and ready for work. He is holding the loop like a pencil, not gripping it like a dagger.

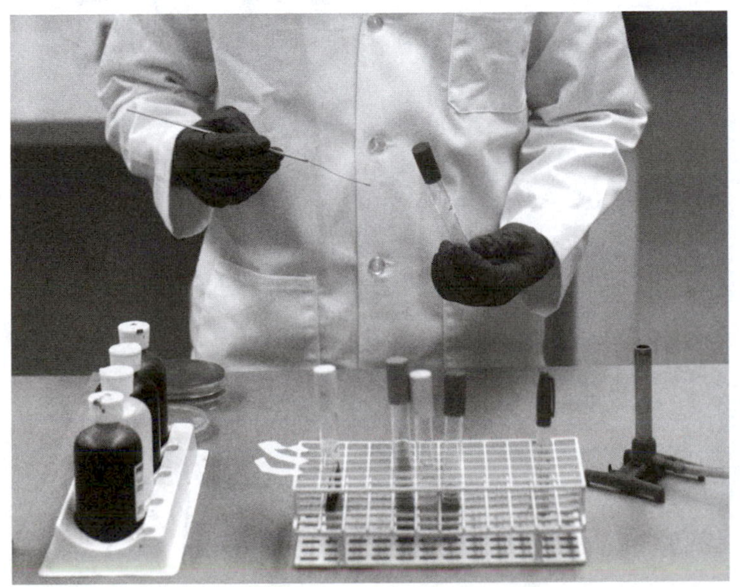

■ FIGURE 1-3 Flaming the Inoculating Loop Wire
Incineration of an inoculating loop's wire is done by passing it through the tip of the flame's inner cone. Begin at the wire's base and continue to the end making sure that all parts are heated to a uniform orange color. Allow the wire to cool before touching it or placing it on/in a culture. The former will burn you; the latter will cause aerosols of microorganisms.

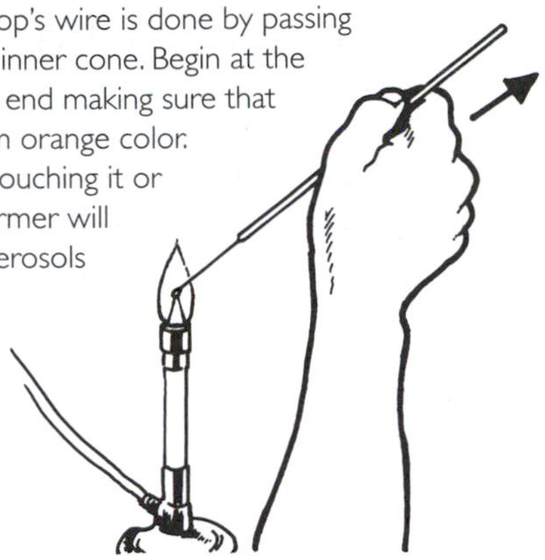

■ FIGURE 1-4 Removing the Cap
Remove the tube's cap with your little finger by pulling the tube away with the other hand; keep your loop hand still. Hold the cap in your little finger during the transfer. When replacing the cap, move the tube back to the cap in order to keep your loop hand still. The replaced cap doesn't need to be on firmly yet—just enough to cover the tube.

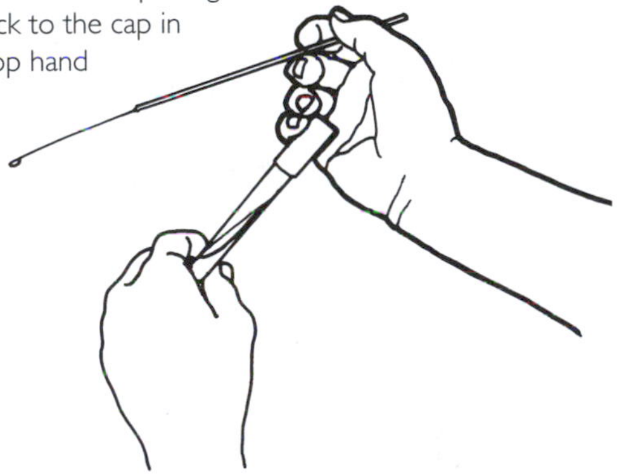

■ FIGURE 1-5 Flaming the Tube
Hold the open tube on an angle to minimize the chance that airborne microbes will drop into it. Quickly pass the tube's mouth through the flame a couple of times. Notice the tube's cap being held in the loop hand.

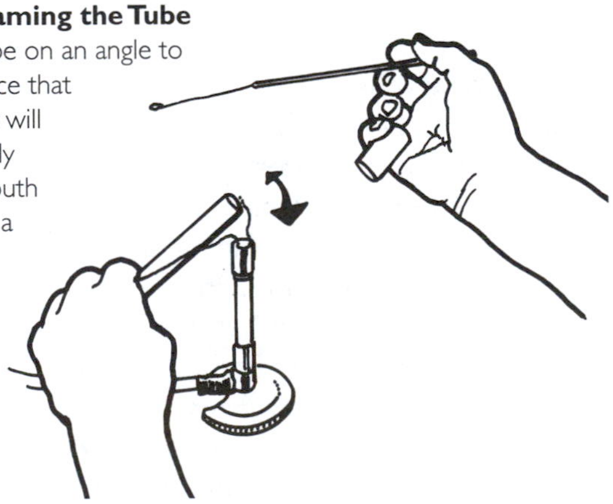

Transfers Using an Inoculating Loop or Needle

Inoculating loops and needles are the most commonly used instruments for transferring microbes between all media types—broths, slants, or plates can be the source, and any can be the destination.

Since loops and needles are handled in the same way, we refer only to loops in the following instructions for ease of reading.

Obtaining a Sample with an Inoculating Loop or Needle

From a Broth

1. Suspend bacteria in the broth with a vortex mixer (Figure 1–6) or by agitating the tube with your fingers (Figure 1–7).
2. Flame the loop (Figure 1-3).
3. Remove and hold the tube's cap with the little finger of your loop hand (Figure 1-4).
4. Flame the open end of the tube by passing it through a flame two or three times (Figure 1-5).
5. Hold the open tube at an angle to prevent airborne contamination.
6. **Holding the loop hand still, move the tube up the wire until the tip is in the broth. Continuing to hold the loop hand still, *remove the tube* from the wire (Figure 1-9). There should be a film of broth in the loop. Be especially careful not to catch the loop tip on the tube lip. This springing action of the loop creates bacterial aerosols.**
7. Flame the tube lip as before. Keep your loop hand still.
8. Keeping the loop hand still (remember, it has growth on it), move the tube to replace its cap (Figure 1-10).
9. What you do next depends on the medium to which you are transferring the growth. Please continue with the appropriate inoculation section.

From a Slant

1. Flame the loop (Figure 1-3).
2. Remove and hold the culture tube's cap with the little finger of your loop hand (Figure 1-4).
3. Flame the open end of the tube by passing it through a flame two or three times (Figure 1-5).
4. With the agar surface facing upward, hold the open tube at an angle to prevent airborne contamination.
5. **Holding the loop hand still, move the tube up the wire until the wire tip is over the desired growth (Figure 1-11). Touch the loop to the growth and obtain the smallest visible mass of bacteria. Then, holding the loop hand still, *remove the tube* from the wire. Be especially careful not to catch the loop tip on the tube lip. This springing action of the loop creates bacterial aerosols.**

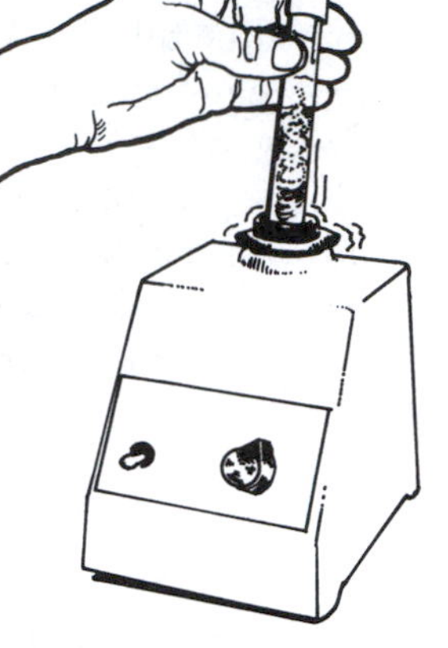

■ FIGURE 1-6 The Vortex Mixer
Bacteria are suspended in a broth with a vortex mixer. The switch on the left has three positions: on (up), off (middle), and touch (down). The rubber boot is only activated when touched if the "touch" position is used; "on" means the boot is constantly vibrating. On the right is a variable speed knob. Caution must be used to prevent broth from getting into the cap or losing control of the tube and causing a spill.

■ FIGURE 1-7 Mixing by Hand
A broth culture should always be mixed prior to transfer. Tapping the tube with your fingers gets the job done safely and without special equipment.

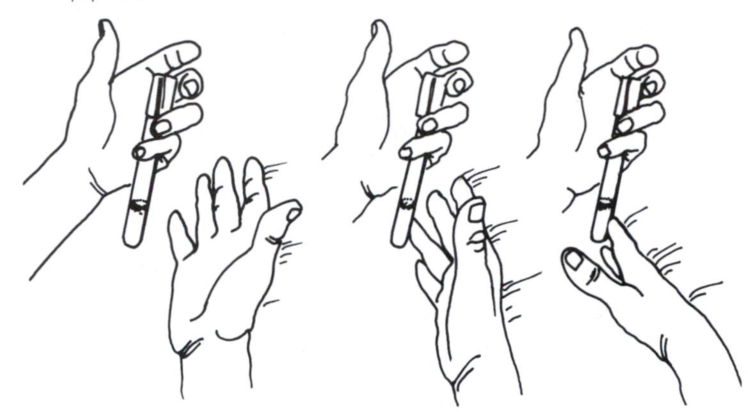

■ FIGURE 1-8 Use the Lid as a Shield
When transferring bacteria to or from a Petri dish, keep the agar surface covered with the lid to minimize airborne contamination.

■ FIGURE 1-9 A Loop and Broth
Hold the open tube on an angle to minimize airborne contamination. When placing a loop into a broth tube or removing it, keep the loop hand still and move the tube. Be careful not to catch the loop on the tube's lip when removing it. This produces aerosols that can be dangerous or produce contamination.

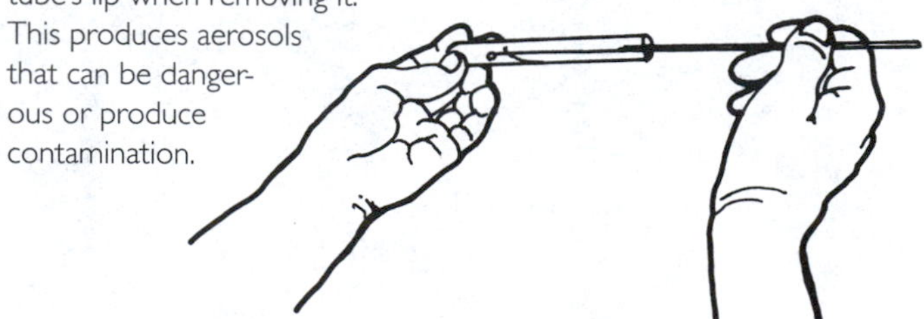

6. Flame the tube lip as before. Keep your loop hand still.
7. Keeping the loop hand still (remember, it has growth on it), move the tube to replace its cap.
8. What you do next depends on the medium to which you are transferring the growth. Please continue with the appropriate inoculation section.

From an Agar Plate

1. Flame the loop (Figure 1-3).
2. Lift the lid of the agar plate, but continue to use it as a cover to prevent contamination from above (Figure 1-8).
3. **Touch the loop to an uninoculated portion of the plate to cool it. (Placing a hot wire on growth may cause spattering of the growth and create aerosols.) Obtain a small amount of bacterial growth by gently touching a colony with the wire tip (Figure 1-8).**
4. Carefully remove the loop from the plate and hold it still as you replace the lid.
5. What you do next depends on the medium to which you are transferring the growth. Please continue with the appropriate inoculation section.

Inoculation of Media with an Inoculating Loop or Needle

Fishtail Inoculation of Agar Slants

Agar slants are generally used for growing stock cultures that can be refrigerated after incubation and maintained for several weeks. Many differential media used in identification of microbes are also slants.

1. Remove the cap of the sterile medium with the little finger of your loop hand and hold it there (Figure 1-4).
2. Flame the tube by quickly passing it through the flame a couple of times. Keep your loop hand still (Figure 1-5).
3. Hold the open tube on an angle to minimize airborne contamination. Keep your loop hand still.
4. **With the agar surface facing upward, carefully move the tube over the wire. Gently touch the loop to the agar surface near the base.**
5. **Beginning at the bottom of the exposed agar surface, drag the loop in a zigzag pattern as the tube is withdrawn (Figure 1-12). Be careful not to cut the agar surface, and be especially careful not to catch the loop tip on the tube lip as you remove it. This springing action of the loop creates bacterial aerosols.**
6. Flame the tube mouth as before. Keep your loop hand still.
7. Keeping the loop hand still (remember, it has growth on it), move the tube to replace its cap.

■ FIGURE 1-10 Replace the Cap
Keeping the loop hand still (remember it has growth on it), move the tube to replace the cap. The cap doesn't have to be on firmly at this point—just enough to cover the tube.

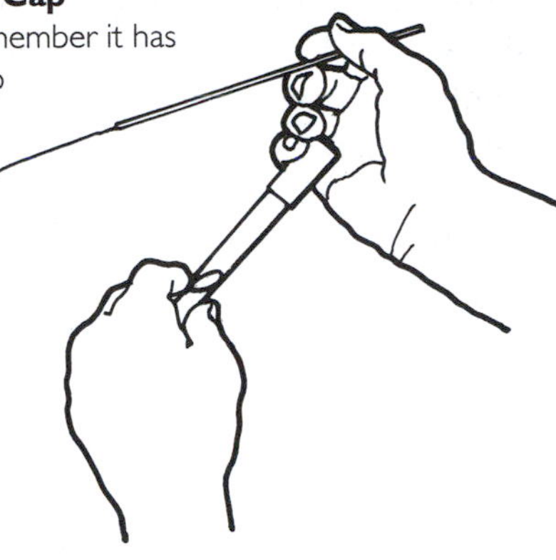

■ FIGURE 1-11 A Loop and an Agar Slant
When placing a loop into a slant tube or removing it, keep the loop hand still and move the tube. Hold the tube so the agar is facing upwards.

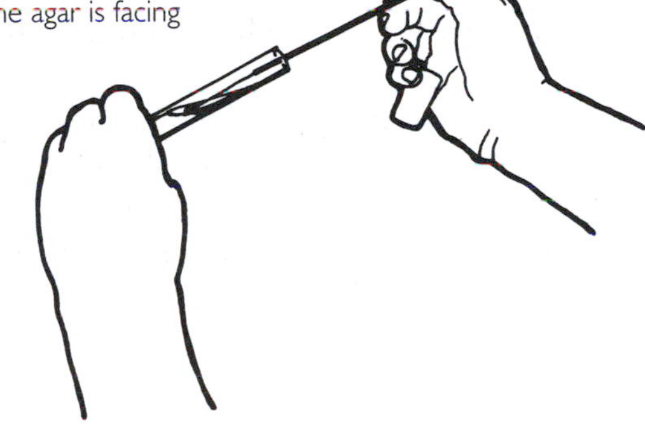

■ FIGURE 1-12 Fishtail Inoculation of a Slant
Begin at the base of the slant surface and gently move the loop back and forth as you withdraw the tube. Be careful not to cut the agar. Sterilize the loop upon completion of the transfer.

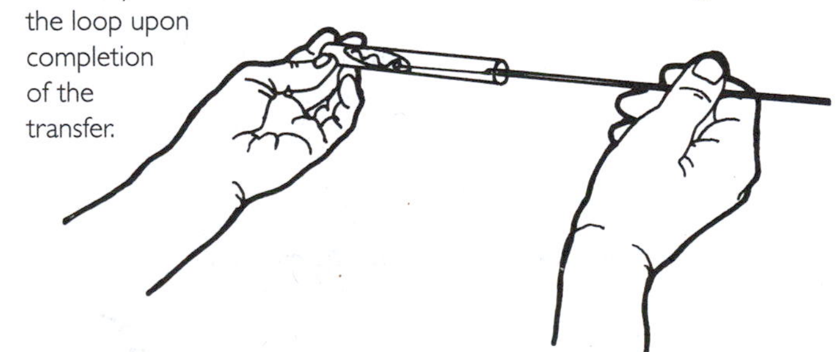

8. Sterilize the loop as before by incinerating it in the Bunsen burner flame. It is especially important to flame it from base to tip now because the loop has lots of bacteria on it.
9. Label the tube with your name, date, and organism. Incubate at the appropriate temperature for the assigned time.

Inoculation of Broth Tubes

Broth cultures are often used to grow cultures for use when fresh cultures or large numbers of cells are desired. Many differential media are also broths.

1. Remove the cap of the sterile medium with the little finger of your loop hand and hold it there (Figure 1-4).
2. Sterilize the tube by quickly passing it through the flame a couple of times. Keep your loop hand still (Figure 1-5).
3. Hold the open tube on an angle to minimize airborne contamination. Keep your loop hand still.
4. **Carefully move the broth tube over the wire (Figure 1-13). Gently swirl the loop in the broth to dislodge microbes.**
5. **Withdraw the tube from over the loop. Before completely removing it, touch the loop tip to the glass to remove any excess broth (Figure 1-14). Then be especially careful not to catch the loop tip on the tube lip when withdrawing it. This springing action of the wire creates bacterial aerosols.**

■ **FIGURE 1-13 Inoculation of a Broth**
When entering or leaving the tube, move the tube and keep the loop hand still. Gently swirl the loop in the broth to transfer the organisms.

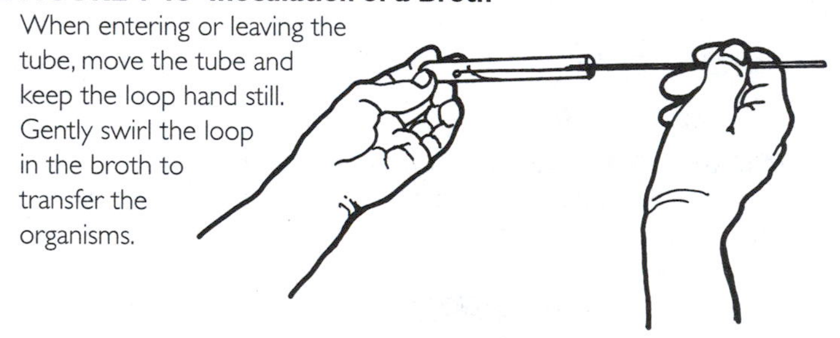

■ **FIGURE 1-14 Remove Excess Broth from Loop**
Before removing it from the tube, touch the loop to the glass to remove excess broth. Failure to do so will result in splattering and aerosols when sterilizing the loop in a flame.

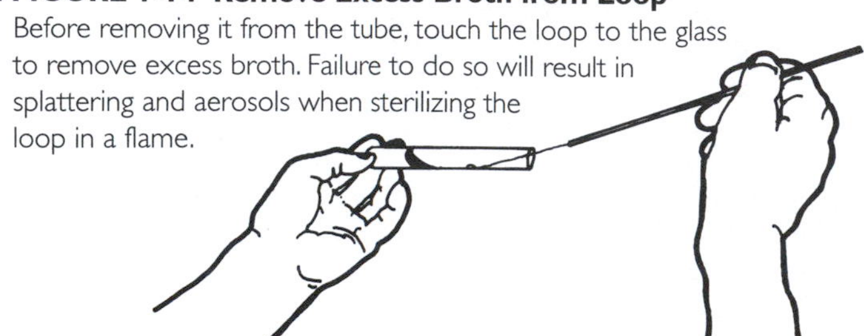

6. Flame the tube lip as before. Keep your loop hand still.
7. Keeping the loop hand still (remember, it has growth on it), move the tube to replace its cap.
8. Sterilize the loop as before by incinerating it in the Bunsen burner flame. It is especially important to flame it from base to tip now because the loop and wire have lots of bacteria on them.
9. Label the tube with your name, date, and organism. Incubate at the appropriate temperature for the assigned time.

Materials Needed For This Exercise

Per Student Group

- Inoculating loop (one per student)
- Five Nutrient Broth tubes
- Five Nutrient Agar slants
- Marking pen and labels
- Vortex mixer (optional)
- Nutrient Agar slant cultures of:
  Bacillus subtilis
  Escherichia coli
  Micrococcus luteus
- Nutrient Broth culture of:
  Staphylococcus epidermidis
- Nutrient Agar Plate culture of:
  Micrococcus roseus

Test Protocol

Lab One

1. Refer to the appropriate section in "Specific Transfer Methods" (pages 5–8) to make the transfers listed.
   a. *B. subtilis* to slant and broth using a loop.
   b. *E. coli* to slant and broth using a loop.
   c. *M. luteus* to slant and broth using a loop.
   d. *S. epidermidis* to slant and broth using an inoculating loop.
   e. *M. roseus* to slant and broth. (For each transfer choose a well-isolated colony and just touch the center with the loop as in Figure 1-8).
2. Label all tubes clearly with your name, the organisms' names, and the date.
3. Incubate *M. luteus* and *M. roseus* at 25°C and the rest at 37°C until next class.

Lab Two

1. Remove the media from the incubators and examine the growth. Record your observations in the table below.
2. Your instructor may ask you to save your media for later use. Otherwise, dispose of all materials in the appropriate autoclave containers.

Observations and Interpretations

Record the appearance of growth on/in each medium. Include the number of apparently different microbial types.

OBSERVATIONS AND INTERPRETATIONS		
Organism	**Medium Inoculated**	
	NA Slant	**NA Broth**
B. subtilis (NA slant)		
E. coli (NA slant)		
M. luteus (NA slant)		
S. epidermidis (NB culture)		
M. roseus (NA plate)		

References

Barkley, W. Emmett and John H. Richardson. 1994. Chapter 29 in *Methods for General and Molecular Bacteriology*. American Society for Microbiology, Washington, DC.

Claus, G. William. 1989. Chapter 2 in *Understanding Microbes—A Laboratory Textbook for Microbiology*. W. H. Freeman and Company, New York, NY.

Darlow, H. M. 1969. Chapter VI in *Methods in Microbiology*, Vol. 1. Edited by J. R. Norris and D. W. Ribbins. Academic Press, Ltd., London.

Fleming, Diane O. 1995. Chapter 13 in *Laboratory Safety—Principles and Practices*, 2nd Ed. Edited by Diane O. Fleming, John H. Richardson, Jerry J. Tulis, and Donald Vesley. American Society for Microbiology, Washington, DC.

Koneman, Elmer W., Stephen D. Allen, William M. Janda, Paul C. Schreckenberger, and Washington C. Winn, Jr. 1997. Chapter 2 in *Color Atlas and Textbook of Diagnostic Microbiology*, 5th Ed. Lippincott-Raven Publishers, Philadelphia, PA.

Murray, Patrick R., Ellen Jo Baron, Michael A. Pfaller, Fred C. Tenover, and Robert H. Yolken. 1995. *Manual of Clinical Microbiology*, 6th Ed. American Society for Microbiology, Washington, DC.

Power, David A. and Peggy J. McCuen. 1988. *Manual of BBL® Products and Laboratory Procedures*, 6th Ed. Becton Dickinson Microbiology Systems, Cockeysville, MD.

1-3 STREAK PLATE METHODS OF ISOLATION

Obtaining isolation of individual bacterial species from a sample is generally the first step in the identification process. A common isolation technique is the **streak plate.** There are many patterns used in streaking an agar plate, the choice of which depends on the source of inoculum and preference of the microbiologist. Streak patterns range from simple to more complex, but all are designed to separate cells deposited on the agar surface and grow isolated colonies. A **quadrant streak** is generally used with samples suspected of high cell density, whereas a simple **zigzag** pattern may be used for samples containing lower cell densities.

Following are descriptions of streak techniques. As in Exercise 1-2, basic skills are printed in normal type, while new skills are printed in **bold.**

Inoculation of Agar Plates Using the Quadrant Streak Method

This inoculation pattern is usually performed as the initial streak for isolation of two or more bacterial species in a mixed culture with suspected high cell density.

1. Obtain the sample of mixed culture with a sterile loop.
2. Lift the lid of the sterile agar plate and use it as a shield to prevent airborne contamination.
3. **Starting at the edge of the plate lightly drag the loop back and forth across the agar surface as shown in Figure 1-15a. Be careful not to cut the agar surface.**
4. **Remove the loop and replace the lid**
5. **Sterilize your loop as before. It is especially important to flame it from base to tip now because the loop has lots of bacteria on it.**
6. **Let the loop cool for a few moments, then perform another streak with the sterile loop beginning at one end of the first streak pattern (Figure 1-15b).**
7. **Sterilize the loop, then repeat with a third streak beginning in the second streak (Figure 1-15c).**
8. **Sterilize the loop, then perform a fourth streak beginning in the third streak and extending into the middle of the plate. Be careful not to enter any streaks but the third (Figure 1-15d).**
9. Sterilize the loop.
10. Label the plate's base with your name, date, and organism(s) inoculated.
11. Incubate the plate in an inverted position for the assigned time at the appropriate temperature.

Zigzag Inoculation of Agar Plates Using a Cotton Swab

This inoculation pattern is usually performed when the sample does not have a high cell density and with pure cultures when isolation is not necessary.

1. Hold the swab comfortably in your dominant hand and lift the lid of the Petri dish with the other. Use the lid as a shield to protect the agar from airborne contamination.
2. **Lightly drag the cotton swab across the agar surface in a zigzag pattern. Be careful not to cut the agar surface (Figure 1-16).**
3. Replace the lid.
4. **Dispose of the swab according to your lab's practices (Figure 1-17).**
5. Label the plate's base with your name, date, and sample.
6. Incubate the plate in an inverted position for the assigned time at the appropriate temperature.

Inoculation of Agar Plates with a Cotton Swab in Preparation for a Quadrant Streak Plate

This inoculation pattern is usually performed as the initial streak for isolation of two or more bacterial species in a mixed culture with suspected high cell density.

1. Hold the swab comfortably in your dominant hand and lift the lid of the Petri dish with the other. Use the lid as a shield to protect the agar from airborne contamination.
2. **Lightly drag the cotton swab back and forth across the agar surface in one quadrant of the plate (Figure 1-18). This replaces the first streak as shown in Figure 1-15a.**
3. **Dispose of the swab according to your lab's practices.**
4. **Further streaking is performed with a loop as described in Figures 1-15b through 1-15d.**
5. Label the plate's base with your name, date, and sample.
6. Incubate the plate in an inverted position for the assigned time at the appropriate temperature.

Photographic Atlas Reference
Streak Plate Method of Isolation Page 13

Materials Needed For This Exercise

Per Student Pair

- Inoculating loop (one per student)
- Four nutrient agar plates
- Two sterile cotton swabs in sterile distilled water
- Four sterile transfer pipettes
- Two sterile microtubes
- Broth cultures of:
  - *Serratia marcescens*
  - *Chromobacterium violaceum*
  - *Staphylococcus aureus*
  - *Staphylococcus epidermidis*

Procedure

Lab One

1. Transfer a few drops of *C. violaceum* and *S. marcescens* to a microtube and mix well. Transfer a loopful of the

mixture to a sterile Nutrient Agar plate and follow the diagrams in Figure 1-15a–d to streak for isolation. Label the plate with your name, the date, and the organisms.
2. Repeat step one with a mixture of *S. aureus* and *S. epidermidis* on a second Nutrient Agar plate. Label the plate with your name, the date, and the organisms.
3. Use the cotton swab to sample an environmental source (see Appendix B), then do a simple zigzag streak on the agar plate (Figure 1-16). Dispose of the swab in an autoclave container. Label the plate with your name, the date, and the sample source.
4. Use the second sterile cotton swab to sample the inside of one partner's cheek or along the gum line of the teeth. Use the swab to perform the first streak of a quadrant streak on the fourth plate (Figure 1-18). Properly dispose of the swab, and then complete the quadrant streak with your loop. Label the plate with your name, the date, and the sample source.
5. Tape the four plates together, invert them, and incubate them at 37°C for 24 to 48 hours.

Lab Two

1. After incubation, examine the plates for isolation.
2. Compare your streak plates with your lab partner's plates and critique each other's technique. Remember, a successful streak plate is one that has isolated colonies; the pattern doesn't have to be textbook quality—it's just that textbook quality provides you with a greater chance of getting isolation.

■ FIGURE 1-15a Beginning the Streak Pattern
Streak the mixed culture back and forth in one quadrant of the agar plate. Use the lid as a shield and do not cut the agar with the loop. Flame the loop, then proceed.

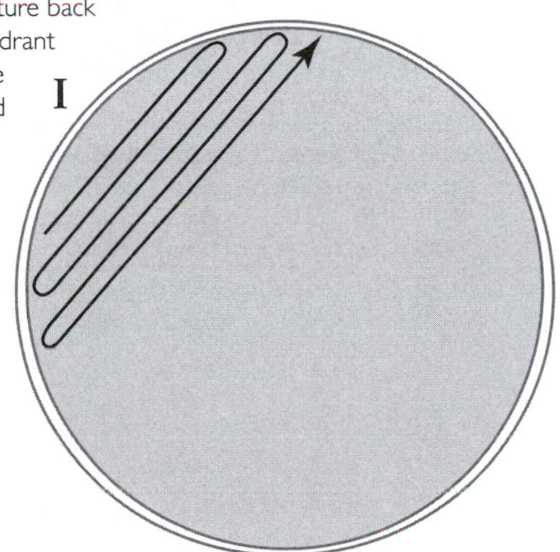

■ FIGURE 1-15c Streak Yet Again
Rotate the plate nearly 90° and streak again using the same wrist motion. Be sure to cool the loop prior to streaking. Flame again.

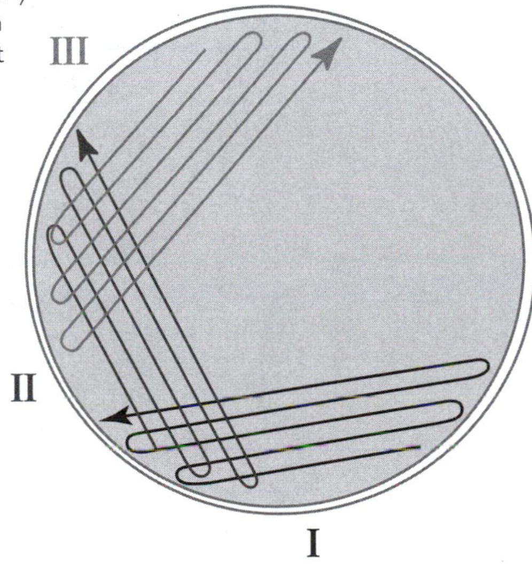

■ FIGURE 1-15b Streak Again
Rotate the plate nearly 90° and touch the agar in an uninoculated region to cool the loop. Streak again using the same wrist motion. Flame the loop.

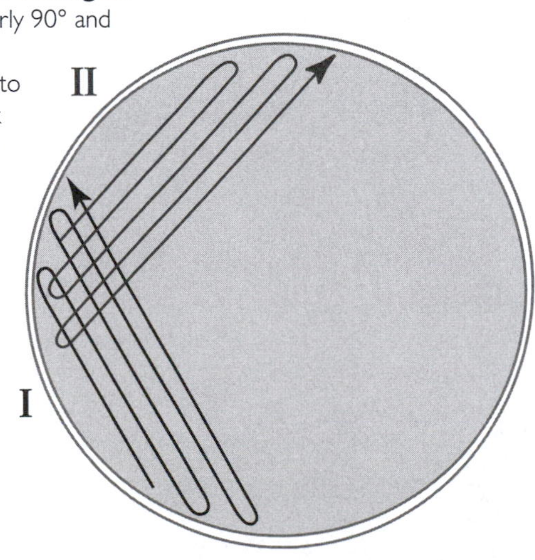

■ FIGURE 1-15d Streak Into the Center
After cooling the loop, streak one last time into the center of the plate. Flame the loop and incubate the plate in an inverted position for the assigned time at the appropriate temperature.

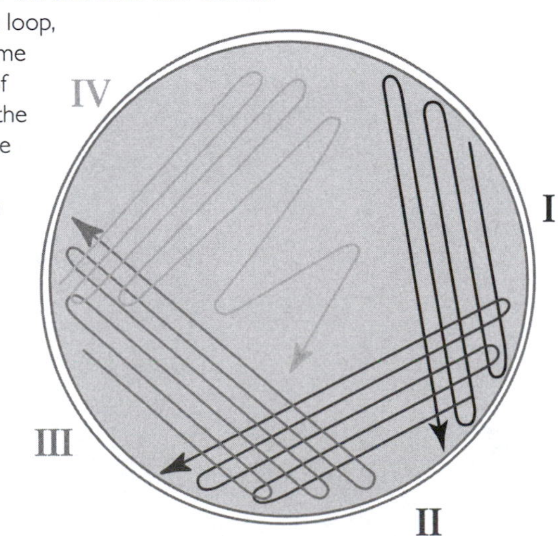

■ FIGURE 1-16 Zigzag Inoculation
Use the cotton swab to streak the agar surface to get isolated colonies after incubation. Be careful not to cut the agar. Properly dispose of the swab in a biohazard container.

■ FIGURE 1-18 Inoculation in Preparation for a Quadrant Streak
If the sample is expected to have a high density of organisms, streak one edge of the plate with the swab. Then continue with the quadrant streak using a loop. Be careful not to cut the agar. Properly dispose of the swab in a biohazard container.

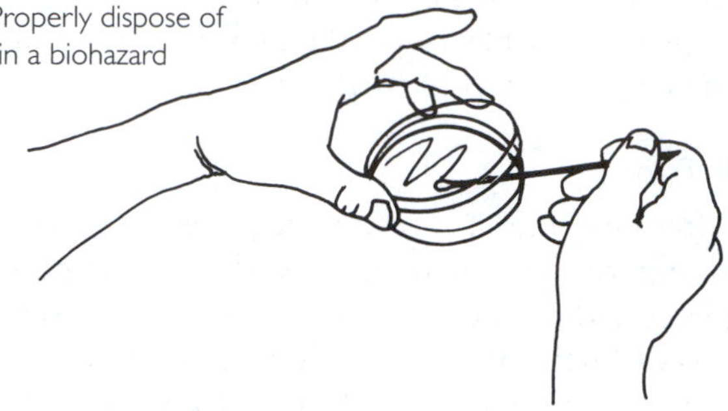

■ FIGURE 1-17 Dispose of the Swab
The swab is contaminated and must be disposed of properly in a biohazard container destined for autoclaving.

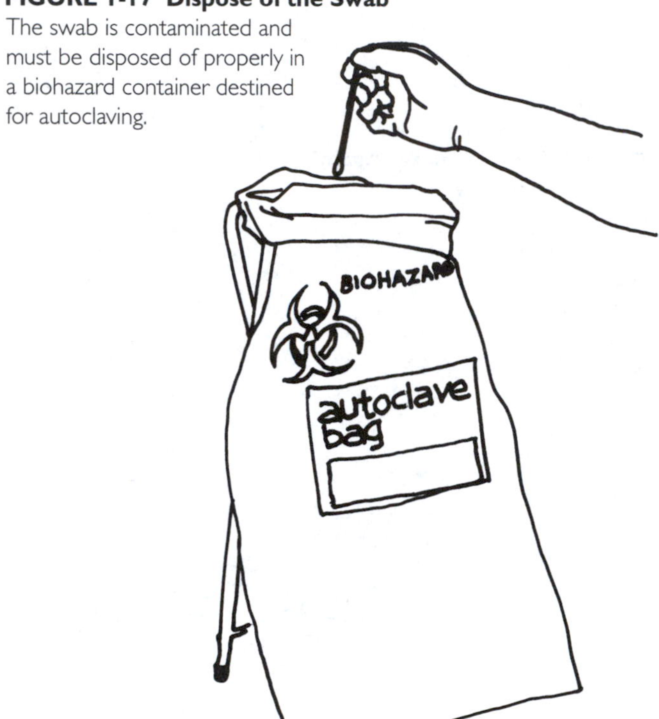

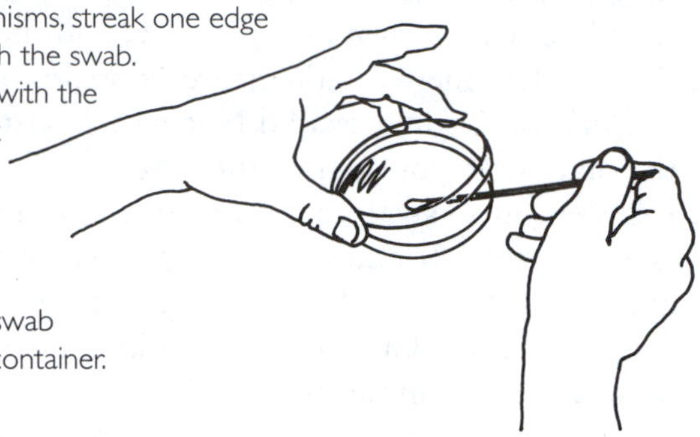

REFERENCES

Collins, C. H. and Patricia M. Lyne. 1995. Chapter 6 in Collins and Lyne's *Microbiological Methods*, 7th Ed. Butterworth-Heineman.

Delost, Maria Dannessa. 1997. Chapter 1 in *Introduction to Diagnostic Microbiology.* Mosby, Inc., St. Louis.

Forbes, Betty A., Daniel F. Sahm, and Alice S. Weissfeld. 2002. Chapter 1 in *Bailey and Scott's Diagnostic Microbiology*, 11th Ed. Mosby-Yearbook, St. Louis, MO.

Koneman, Elmer W., Stephen D. Allen, William M. Janda, Paul C. Schreckenberger, and Washington C. Winn, Jr. 1997. Chapter 2 in *Color Atlas and Textbook of Diagnostic Microbiology*, 5th Ed. J. B. Lippincott Company, Philadelphia, PA.

Power, David A. and Peggy J. McCuen. 1988. Pages 2 and 3 in *Manual of BBL® Products and Laboratory Procedures*, 6th Ed. Becton Dickinson Microbiology Systems, Cockeysville, MD.

1-4 SPREAD PLATE METHOD OF ISOLATION

The spread plate technique is a method of isolation in which a diluted microbial sample is deposited on an agar plate and uniformly spread across the surface with a glass rod. After incubation, given the proper dilution, the plate is expected to produce individual colonies that can be transferred to start a pure culture.

Following is a description of the spread plate technique. As in Exercise 1-2, basic skills are printed in normal type, while new skills are printed in **bold**.

The Spread Plate Technique

1. **Arrange the alcohol beaker, Bunsen burner and agar plate as shown in Figure 1-19. This arrangement minimizes the chances of catching the alcohol on fire.**
2. Lift the plate's lid and use it as a shield to protect from airborne contamination.
3. **Using an appropriate pipette, deposit the designated volume on the agar surface. (Please see Appendices C and D for pipette use.) From this point, the remainder of steps should be completed within about 15 seconds to prevent the inoculum from soaking into the agar.**
4. **The pipetting instrument used to inoculate the medium is contaminated, and must be disposed of properly.** Each lab has its own specific procedures and your instructor will advise you what to do.
5. **Remove the glass spreading rod from the alcohol and pass it through the flame to ignite the alcohol (Figure 1-20). Remove the rod from the flame and allow the alcohol to burn off completely. Do not leave the rod in the flame; the combination of the alcohol and brief flaming are sufficient to sterilize it.** ***Be careful not to drop any flaming alcohol on the work surface. Be especially careful not to drop flaming alcohol back into the alcohol beaker.***
6. **After the flame has gone out on the glass rod, lift the lid of the plate and use it as a shield from airborne contamination. Then, touch the rod to the agar surface away from the inoculum in order to cool it.**
7. **To spread the inoculum, hold the plate lid with the base of your thumb and index finger, and use the tip of your thumb and middle finger to rotate the base (Figure 1-21). At the same time, move the rod in a back-and-forth motion across the agar surface. After a couple of turns, do one last turn with the rod next to the plate's edge. Alternatively, the plate may be placed on a rotating platform and inoculated.**
8. **Remove the rod from the plate and replace the lid.**
9. **Return the rod to the alcohol in preparation for the next inoculation. There is no need to flame it again.**
10. Label the plate base with your name, date, organism, and any other relevant information.
11. Incubate the plate in an inverted position at the appropriate temperature for the assigned time. (If you plated a large volume of inoculum, wait a few minutes and allow it to soak in before inverting the plate.)

■ FIGURE 1-19 Safety First
The spread plate technique requires a Bunsen burner, a beaker with alcohol, a glass spreading rod, and the plate. Position these components in your work area as shown: isopropyl alcohol, flame, and plate. This arrangement reduces the chance of accidentally catching the alcohol on fire.

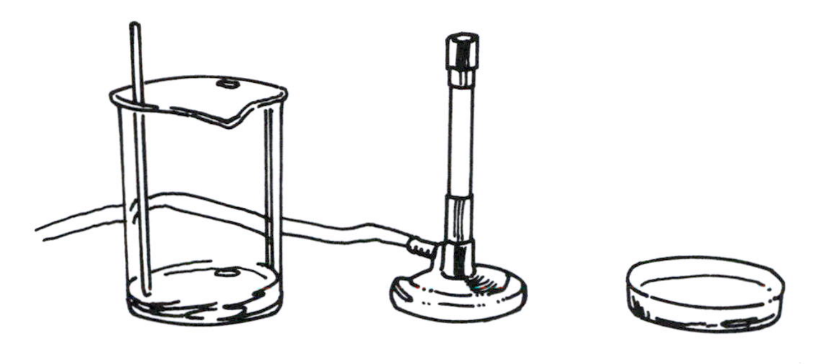

■ FIGURE 1-20 Sterilize the Glass Rod
Remove the glass spreading rod from the alcohol and pass it through the flame to ignite the alcohol. Remove the rod from the flame and allow the alcohol to burn off completely. Do not leave the rod in the flame; the combination of the alcohol and brief flaming are sufficient to sterilize it. *Be careful not to drop any flaming alcohol on the work surface.*

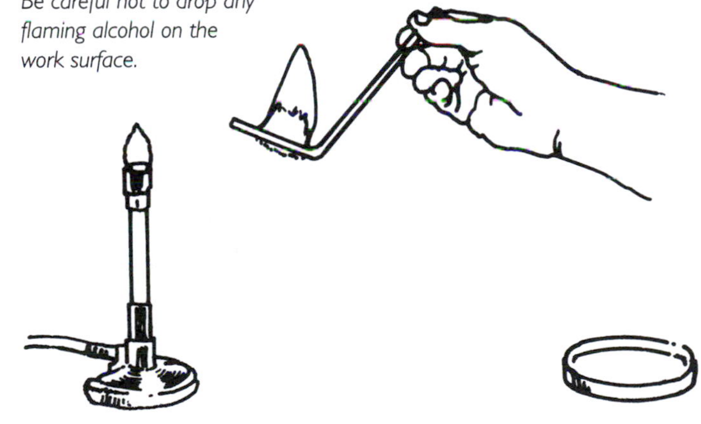

■ FIGURE 1-21 Spread the Inoculum
After the flame has gone out on the rod, lift the lid of the plate and use it as a shield from airborne contamination. Then, touch the rod to the agar surface away from the inoculum in order to cool it. To spread the inoculum, hold the plate lid with the base of your thumb and index finger, and use the tip of your thumb and middle finger to rotate the base. At the same time, move the rod in a back-and-forth motion across the agar surface. After a couple of turns, do one last turn with the rod next to the plate's edge.

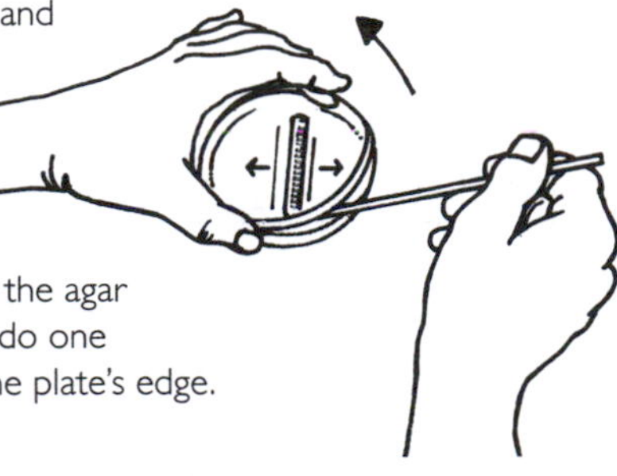

Photographic Atlas Reference
Spread Plate Method of Isolation Page 83

MATERIALS NEEDED FOR THIS EXERCISE

Per Student Pair

- Inoculating loop (each student)
- Six sterile plastic transfer pipettes
- 500 mL beaker with 50 mL of isopropyl alcohol
- Glass spreading rod
- Bunsen burner
- Striker
- Four Nutrient Agar plates
- One sterile microtube
- Four capped microtubes with about 1 mL sterile dH_2O
- Broth cultures of:
  - *Escherichia coli*
  - *Serratia marcescens*

PROCEDURE

Lab One

1. Using a different pipette for each, transfer a few drops of *E. coli* and *S. marcescens* to a microtube. Cap the tube and mix well with a vortex mixer. Or, use the second pipette to mix well by gently drawing and dispensing the mixture in and out of the tube a couple of times. Do not spray the mixture!
2. Label the four microtubes containing sterile dH_2O "A", "B", "C", and "D".
3. Label the four Nutrient Agar plates "A", "B", "C", and "D".
4. Transfer a loopful of the mixture to Tube A and mix well with the loop or a vortex mixer. If a vortex mixer is used, be sure to cap the tube.*
5. Transfer a loopful of the mixture in Tube A to Tube B and mix well with the loop or a vortex mixer.
6. Transfer a loopful of the mixture in Tube B to Tube C and mix well with the loop or a vortex mixer.
7. Transfer a loopful of the mixture in Tube C to Tube D and mix well with the loop or a vortex mixer.
8. Using a sterile transfer pipette, place a couple of drops of sample from Tube A on Plate A. Spread the inoculum with a glass rod as described in Figure 1-21. Let the plate sit for a few minutes.
9. Repeat Step 8 for Tubes B, C, and D and Plates B, C, and D, respectively.
10. Tape the four plates together, invert them, and incubate them at 25°C for 24 to 48 hours.

* Since this is not a quantitative procedure, it is not necessary flame the loop between transfers.

Lab Two

1. After incubation, examine the plates for isolation. *S. marcescens* produces reddish colonies and *E. coli* produces buff-colored colonies.
2. Fill in the table provided.

REFERENCES

Clesceri, WEF, Chair; Arnold E. Greenberg, APHA; Andrew D. Eaton, AWWA ; and Mary Ann H. Franson. 1998. Pages 9–38 in *Standard Methods for the Examination of Water and Wastewater*, 20th Edition. Joint publication of American Public Health Association, American Water Works Association and Water Environment Federation. APHA Publication Office, Washington, DC.

Downes, Frances Pouch and Keith Ito. 2001. Page 57 in *Compendium of Methods for the Microbiological Examination of Foods*. American Public Health Association. Washington DC.

Gerhard, Philipp, R. G. E. Murray, Willis A. Wood, and Noel R. Kreig. 1994. Pages 255–257 in *Methods for General and Molecular Bacteriology*. American Society for Microbiology, Washington, DC.

■ OBSERVATIONS AND INTERPRETATIONS

Record your observations in the table below.

OBSERVATIONS AND INTERPRETATIONS		
ORGANISM	**PLATE(S) WITH ISOLATION**	**COMMENTS**
E. coli		
S. marcescens		

SECTION TWO

Microbial Growth

2

Microorganisms are extraordinarily diverse. Every species demonstrates unique characteristics (some readily observable, some not) defined in part by the environment in which it lives, its metabolic requirements and abilities, and its resistance or susceptibility to control mechanisms. Those characteristics and the factors that affect them are what this section is about.

You will begin this section by sensitizing yourself to the ubiquitous nature of microorganisms. You will then examine some microbial growth characteristics and cultivation methods. Next you will look at some environmental factors affecting microbial growth, including pH, oxygen, temperature, and osmotic pressure. Finally, you will examine the resistance and susceptibility of microorganisms to antiseptics, disinfectants, and antibiotics.

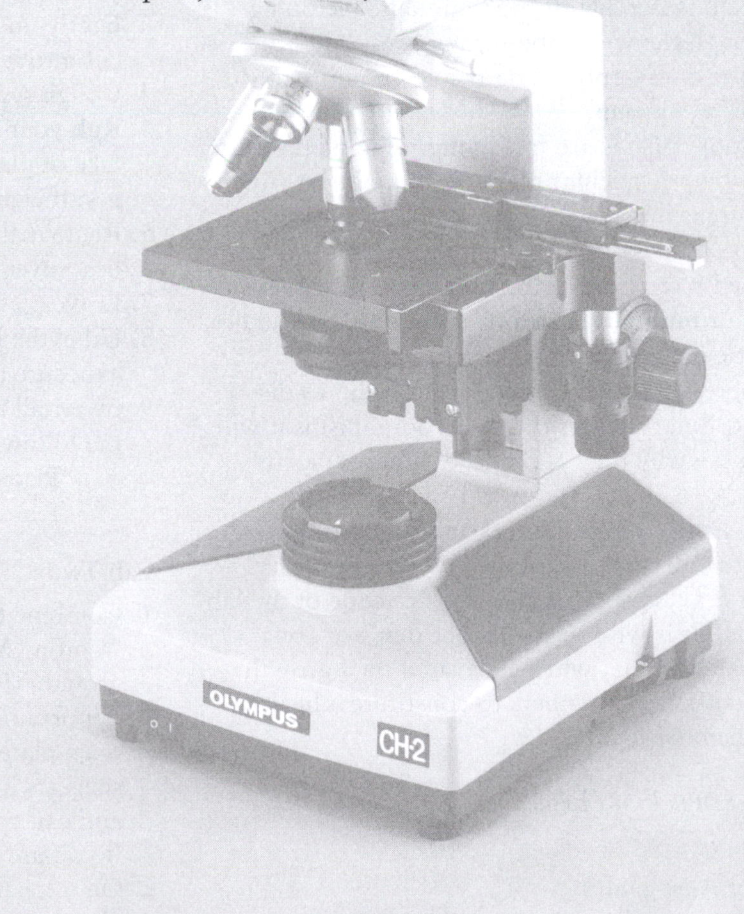

Diversity and Ubiquity of Microorganisms

Microorganisms are everywhere there is life. They can be found in water, in soil, and even in air. They inhabit internal and external surfaces of other organisms. They are around us all the time, yet most people are scarcely aware of their presence. That awareness is about to change for you. In this section, you will grow microorganisms from seemingly uninhabited sources. You will then learn to identify the various types of growth characteristics produced by these "invisible" co-habitants when cultivated in broth and on solid media.

A microbiologist's typical first step toward identification of an organism is a visual examination of its growth. Because bacteria produce such a great variety of cultural characteristics, much can be determined by simply looking at the colonies produced when grown (under controlled conditions) on an agar plate, especially if the sample's source is known.

Distinguishing different growth patterns is a skill that you will learn in this section and will employ as you progress through the semester. Note the growth characteristics of all the organisms provided for your laboratory exercises and jot them down. When the time comes to identify your unknown species, you will find your good record keeping a valuable asset!

2-1 UBIQUITY OF MICROORGANISMS

A phrase frequently encountered when researching literature on microorganisms is "ubiquitous in nature." This means that the organism being considered can be found just about everywhere. More specifically, it could likely be isolated from soil, water, plants, and animals (including humans). Although the word ubiquitous doesn't apply to every species it does apply to many and certainly to microorganisms as a group.

Many microorganisms are **free living.** That is, they do not reside on or in a particular plant or animal **host**. Any area including areas outside the typical host organism where a microbe resides and serves as a potential source of infection is called a **reservoir**. Most microorganisms, even the **commensal** or **mutualistic** strains inhabiting our bodies, are **opportunistic pathogens**.

As you progress through this class and begin to discover the extraordinary diversity in microorganisms it will become easy to see why they are indeed "ubiquitous in nature."

In today's exercise you will work in small groups to sample and culture several locations in your laboratory. Your instructor may have other locations outside of the lab to sample as well. It is wise to remember that even relatively "harmless" bacteria, when cultivated on a growth medium, are in sufficient numbers to constitute a health hazard. Treat them with care.

MATERIAL NEEDED FOR THIS EXERCISE

Per Student Group

- Eight Nutrient Agar plates
- One sterile cotton swab

PROCEDURE

Lab One

1. Number the plates 1 through 8.
2. Open plate number 1 and expose it to the air for 30 minutes or longer.
3. Use the cotton swab to sample your desk area and then streak plates 2 and 3 as in Figure 2-1. Press very lightly on the agar as you streak it. You don't want to cut into it.
4. Cough several times on the agar surface of plate 4.
5. Rub your hands together and then touch the agar surface of plate 5 lightly with your fingertips. A light touch is sufficient; touching too firmly will crack the agar.
6. Remove the lid of plate 6 and vigorously scratch your head over it.
7. Leave plates 7 and 8 covered; do not open them.
8. Label the base of each plate with the date, type of exposure it has received, and your group's name.
9. Invert all plates and incubate them for 24 to 48 hours at the following temperatures:
   Plates 1, 2, and 8—25°C
   Plates 3, 4, 5, 6, and 7—37°C

Lab Two

1. Combine these plates with those in Exercise 2-2 "Colony Morphology" to examine and compare the growth. (Remember that most microorganisms are opportunistic pathogens; handle them carefully. Do not open plates containing fuzzy growth; a fuzzy appearance suggests fungal growth with spores that are spread easily and can contaminate the laboratory and other cultures. If you are in doubt, check with your instructor.)
2. On a separate piece of paper, draw representative examples of growth from each of your plates. Be sure to label them according to incubation time, temperature, and source of inoculum.

■ **FIGURE 2-1 Simple Streak Pattern on Nutrient Agar**

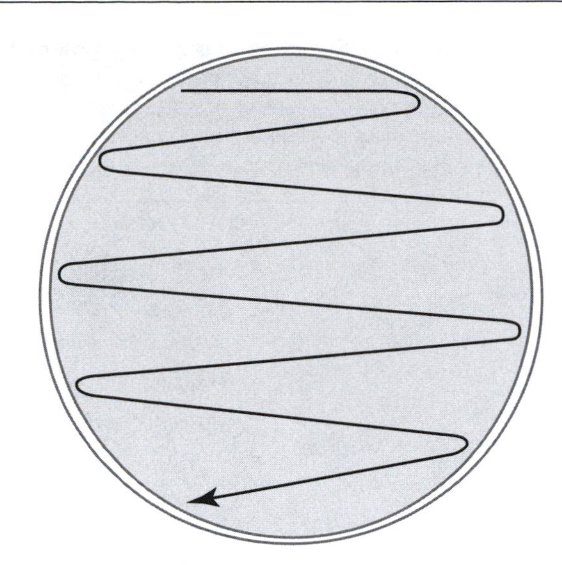

References

Forbes, Betty A., Daniel F. Sahm, Alice S. Weissfeld. 2002. Chapter 10 in *Bailey & Scott's Diagnostic Microbiology*, 11th Ed. Mosby, Inc., St. Louis, MO.

Holt, John G. (Editor). 1994. *Bergey's Manual of Determinative Bacteriology*, 9th Ed. Williams and Wilkins, Baltimore, MD.

Koneman, Elmer W., Stephen D. Allen, William M. Janda, Paul C. Schreckenberger, and Washington C. Winn, Jr. 1997. *Color Atlas and Textbook of Diagnostic Microbiology*, 5th Ed. J. B. Lippincott Company.

Varnam, Alan H. and Malcolm G. Evans. 2000. *Environmental Microbiology.* ASM Press, Washington, DC.

2-2 COLONY MORPHOLOGY

Photographic Atlas Reference
Colony Morphology Page 1

MATERIAL NEEDED FOR THIS EXERCISE

Per Student Group

- Colony counter (optional)
- Metric ruler
- Streak plate cultures of:
  - *Chromobacterium violaceum*
  - *Micrococcus roseus*
  - *Bacillus subtilis*
  - *Serratia marcescens*

PROCEDURE

1. Using the terms in Figure 2-2, describe the colonies in today's cultures and the colonies on the plates from Exercise 2-1. Measure colony diameters with a ruler and include them with your descriptions in the table below. It may be helpful to use a colony counter.
2. Save the four plates from today's exercise for Exercise 2-3, "Growth Patterns in Broth." Discard the plates from Exercise 2-1 in an appropriate autoclave container.

FIGURE 2-2 A Sampling of Bacterial Colony Features
Use these terms to describe colonial morphology. Color, surface characteristics (dull or shiny), consistency (dry, butyrous-buttery, or moist) and optical properties (opaque or translucent), should also be included in colony morphology descriptions.

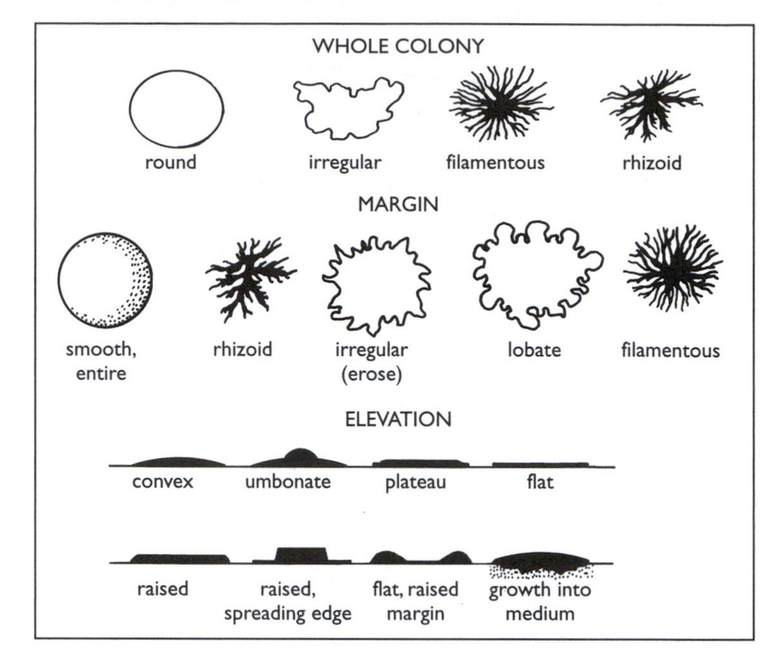

REFERENCES

Claus, G. William. 1989. Chapter 14 in *Understanding Microbes—A Laboratory Textbook for Microbiology*. W. H. Freeman and Company, New York, NY.

Collins, C. H., Patricia M. Lyne and J. M. Grange. 1995. Chapter 6 in *Collins and Lyne's Microbiological Methods*, 7th Ed. Butterworth-Heineman, Oxford.

Forbes, Betty A., Daniel F. Sahm, and Alice S. Weissfeld. 2002. Chapter 10 in *Bailey and Scott's Diagnostic Microbiology*, 11th Ed. Mosby-Yearbook, St. Louis, MO.

Koneman, Elmer W., Stephen D. Allen, William M. Janda, Paul C. Schreckenberger, and Washington C. Winn, Jr. 1997. Chapter 2 in *Color Atlas and Textbook of Diagnostic Microbiology*, 5th Ed. J. B. Lippincott Company, Philadelphia, PA.

OBSERVATIONS AND INTERPRETATIONS

Record your observations in the table below.

OBSERVATIONS AND INTERPRETATIONS	
ORGANISM/PLATE	**COLONY DESCRIPTION**

2-3 Growth Patterns in Broth

Photographic Atlas Reference
Growth in Broth Page 9

Material Needed For This Exercise

Per Student Group

- Five Nutrient Broth tubes
- Your plates from Exercise 2-2

Procedure

Lab One

1. From each plate saved in Exercise 2-2, aseptically transfer growth from a single colony to a Nutrient Broth. Make one transfer per plate until you have inoculated all broths. Leave the fifth broth uninoculated. It will be used as a control tube for comparison. Refer to Section 1 for help with transfers.
2. Label each tube with your name, the date, the medium, and the name of the organism.
3. Incubate the *B. subtilis* broth at 37°C and all other broths (including the uninoculated control) at 25°C for 24 to 48 hours.

Lab Two

1. Examine the tubes and describe the different growth patterns in the table below. Refer to Figure 2-2 in the Photographic Atlas as needed.

References

Claus, G. William. 1989. Chapter 17 in *Understanding Microbes—A Laboratory Textbook for Microbiology*. W. H. Freeman and Company, New York, NY.

Forbes, Betty A., Daniel F. Sahm, and Alice S. Weissfeld. 2002. Chapter 10 in *Bailey and Scott's Diagnostic Microbiology*, 11th Ed. Mosby-Yearbook, St. Louis, MO.

■ Observations and Interpretations

Record your observations in the table below.

Observations and Interpretations	
Organism	**Description of Growth**
Uninoculated Broth	

2-4 GROWTH PATTERNS ON SLANTS

Photographic Atlas Reference
Growth Patterns on Agar Slants Page 8

MATERIALS NEEDED FOR THIS EXERCISE

Per Student Group

- Five Nutrient Agar slants
- Nutrient Broth cultures of:
  - *Micrococcus luteus*
  - *Staphylococcus epidermidis*
  - *Lactobacillus plantarum*
  - *Mycobacterium smegmatis*
  - *Bacillus subtilis*

PROCEDURES

Lab One

1. Aseptically transfer one loopful of each broth culture to an agar slant. Refer to Section 1 for help with transfers. For best results inoculate each slant with a single straight streak. Uniform inoculations like this will help enable comparison of true growth characteristics. Leave the sixth tube uninoculated for later comparison and media quality control.
2. Label each tube with your name, the date, the medium, and the name of the organism.
3. Incubate *M. luteus* and the control slant at 25°C and all other slants at 37°C for 24 to 48 hours.

Lab Two

1. Examine the tubes and describe the different growth patterns in the table provided. Include a sketch of a representative portion of each. Refer to Figure 2-2 and the Photographic Atlas as needed.

REFERENCE

Claus, G. William. 1989. Chapter 17 in *Understanding Microbes—A Laboratory Textbook for Microbiology*. W. H. Freeman and Company, New York, NY.

■ OBSERVATIONS AND INTERPRETATIONS

Record your observations in the table below.

OBSERVATIONS AND INTERPRETATIONS	
ORGANISM	**DESCRIPTION OF GROWTH**
Uninoculated Control	

Environmental Factors Affecting Microbial Growth

Bacteria have very limited control over their internal environments. Whereas many eukaryotes have evolved sophisticated internal control mechanisms, bacteria are almost completely dependent upon external factors to provide suitable conditions for their existence. Minor environmental changes can dramatically change a microorganism's ability to transport materials across the membrane, perform complex enzymatic reactions, and maintain critical cytoplasmic pressure.

One way to observe bacterial responses to environmental changes is to artificially manipulate external factors and measure growth rate. It should be understood that growth rate, as it is used in this manual, is synonymous with reproductive rate. Optimal growth conditions, therefore, result in faster growth and greater cell density (turbidity) than do less than optimal conditions.

In this series of laboratory exercises you will examine the effects of oxygen, temperature, pH, osmotic pressure, and nutritional resource availability on bacterial growth rate. You will also learn some methods for cultivating anaerobic bacteria.

2-5 EVALUATION OF MEDIA

Living things are composed of compounds from four biochemical families: proteins, carbohydrates, lipids, and nucleic acids. Whereas all organisms share this fundamental chemical composition, they differ greatly in their abilities to make these molecules. Some are capable of making them out of a simple carbon compound, carbon dioxide (CO_2). These organisms are called **autotrophs** and require the least "assistance" from the environment to grow. The remaining organisms, called **heterotrophs,** require preformed organic compounds from the environment. Some heterotrophs are metabolically flexible and require only a few simple organic compounds from which to make all their biochemicals. Others require a greater portion of their organic compounds from the environment. An organism that relies greatly on the environment to supply ready-made organic compounds is referred to as **fastidious**. Among the heterotrophs there is a wide range of fastidiousness—from very fastidious to the **nonfastidious**. Autotrophs are less fastidious than the most nonfastidious heterotroph.

Successful cultivation of microbes in the laboratory requires an ability to satisfy their nutritional needs. The absence of a single required chemical resource prevents growth. In general, the more fastidious the organism, the more ingredients a medium must have. **Undefined** media are composed of extracts from plant or animal sources and are very rich in nutrients. Even though the exact composition of the medium and the amount of each ingredient are unknown, undefined media are useful in growing the greatest variety of microbes. A **defined** medium is one in which the amount and identity of every ingredient is known. Defined media typically support a narrower range of organisms.

The ability of a microbiologist to cultivate microorganisms requires some knowledge of their metabolic needs. One quick way to make such a determination is to transfer an organism to media containing different nutritional components and observe how well it grows. In this exercise you will evaluate the ability of four media—Nutrient Broth, Glucose Broth, Yeast Extract Broth, and Glucose Salts Broth—to support bacterial growth.

MATERIAL NEEDED FOR THIS EXERCISE

Per Student Group

- 5 tubes each of:
  - Nutrient Broth
  - Glucose Broth
  - Yeast Extract Broth
  - Glucose Salts Medium
- Nutrient Broth cultures of:
  - *Escherichia coli*
  - *Staphylococcus epidermidis*
  - *Lactococcus lactis*
  - *Moraxella catarrhalis*

Media Recipe

Nutrient Broth

Beef Extract	3.0 g
Peptone	5.0 g
Distilled or deionized water	1.0 L

Glucose Broth

Peptone	10.0 g
Glucose	5.0 g
NaCl	5.0 g
Distilled or deionized water	1.0 L

Yeast Extract Broth

NaCl	5.0 g
Yeast Extract	5.0 g
Nutrient Broth	1.0 L

Glucose Salts Medium

Glucose	5.0 g
NaCl	5.0 g
$MgSO_4$	0.2 g
$(NH_4)H_2PO_4$	1.0 g
K_2HPO_4	1.0 g
Distilled or deionized water	1.0 L

PROCEDURE

Lab One

1. Inoculate each of the media with a single loopful of broth culture. Leave one of each tube uninoculated.
2. Incubate all tubes (including the uninoculated ones) at 37°C for 24–48 hours.

Lab Two

1. Examine the tubes for turbidity. Score relative amounts of growth using "0" for no growth, and "+", "++", and "+++" for successively greater degrees of growth. (Note: record what you see. Do not expect one of each result just because there are four organisms.)
2. Record your results and interpretations in the table provided.

REFERENCES

Delost, Maria Dannessa. 1997. Page 144 in *Introduction to Diagnostic Microbiology. A Text and Workbook*. Mosby, Inc., St. Louis, MO.

DIFCO Laboratories. 1984. *DIFCO Manual,* 10th Ed. DIFCO Laboratories, Detroit, MI.

Power, David A. and Peggy J. McCuen. 1988. *Manual of BBL® Products and Laboratory Procedures*, 6th Ed. Becton Dickinson Microbiology Systems, Cockeysville, MD.

■ OBSERVATIONS AND INTERPRETATIONS

Record your observations in the table below.

OBSERVATIONS AND INTERPRETATIONS

ORGANISM	NUTRIENT BROTH	GLUCOSE BROTH	YEAST EXTRACT BROTH	GLUCOSE SALTS MEDIUM	INTERPRETATION (RELATIVE FASTIDIOUSNESS)
Uninoculated Broth					

Aerotolerance

Aerotolerance is the term used to indicate the ability of a microbe to grow in the presence of air, or more specifically, oxygen. This is a fundamental characteristic of bacteria that must be considered for cultivation purposes, but may also be used for identification. *Bergey's Manual of Determinative Bacteriology, 9th Ed.* includes aerotolerance category in many of its section titles. Aerotolerance categories include **obligate (strict) aerobes, facultative anaerobes, aerotolerant anaerobes, microaerophiles,** and **obligate (strict) anaerobes.** This exercise introduces you to a sampling of these groups and means used for distinguishing them.

2-6 AGAR DEEP STAB

Photographic Atlas Reference
Aerotolerance Page 9

MATERIAL NEEDED FOR THIS EXERCISE

Per Student Group

- Five Agar Deep Stab tubes
- Inoculating needle
- Fresh cultures of:
  - *Clostridium sporogenes*
  - *Pseudomonas aeruginosa*
  - *Rhodospirillum rubrum*
  - *Staphylococcus aureus*

PROCEDURE

Lab One

1. With a heavy inoculum on your inoculating needle, carefully stab the agar tubes with the test organisms. Refer to Appendix B, Figure B-3 for correct stab technique. Note: Try not to move the needle sideways as you insert and remove it from the agar. This precaution will minimize the introduction of oxygen to the medium.
2. Stab the fifth tube with your sterile needle.
3. Label each tube with your name, the date, and the medium.
4. Incubate the tubes at 37°C for 24 to 48 hours.

Lab Two

1. Examine the tubes and determine the aerotolerance category of each organism.
2. Record the results in the table provided.

REFERENCES

Baron, Ellen Jo, Lance R. Peterson, and Sydney M. Finegold. 1994. Chapter 9 in *Bailey and Scott's Diagnostic Microbiology*, 9th Ed. Mosby-Yearbook, Inc. St. Louis, MO.

Koneman, Elmer W., Stephen D. Allen, William M. Janda, Paul C. Schreckenberger, and Washington C. Winn, Jr. 1997. Chapter 14 in *Color Atlas and Textbook of Diagnostic Microbiology*, 5th Ed. J. B. Lippincott Company, Philadelphia, PA.

■ OBSERVATIONS AND INTERPRETATIONS

Record your observations in the table below.

DATA AND OBSERVATIONS

ORGANISM	REGION OF GROWTH	AEROTOLERANCE CATEGORY
Sterile Stab		

2-7 AGAR SHAKES

Photographic Atlas Reference
Aerotolerance Page 9

MATERIALS NEEDED FOR THIS EXERCISE

Per Student Group

- Five Agar Shake tubes (liquefied and held in a 50°C water bath)
- Four Sterile transfer pipettes
- Fresh broth cultures of:
  - *Clostridium sporogenes*
  - *Pseudomonas aeruginosa*
  - *Rhodospirillum rubrum*
  - *Staphylococcus aureus*

PROCEDURE

Lab One

1. Prepare five labels, one for each organism and one control. Because it is important to leave the liquid agar in the water bath until it is needed, place the labels on the tubes one at a time as you obtain them.
2. Remove one liquid agar from the water bath. (Check to see that the temperature of the water bath is 50°C. If the temperature is too high, it may kill the organism; if it is too low, it will solidify before you inoculate it.) Using a sterile pipette transfer a heavy inoculum to the liquefied agar. Immediately mix it thoroughly. Place the tube in a rack and allow it to solidify.
3. One at a time, do the same with the other three organisms. Thoroughly mix the control tube and allow it to solidify uninoculated.
4. When all agar tubes have solidified incubate them at 37°C for 24 to 48 hours.

Lab Two

1. After incubation, examine the tubes and determine the aerotolerance category of each organism.
2. Record your observations and interpretations in the table provided.

REFERENCES

Baron, Ellen Jo, Lance R. Peterson, and Sydney M. Finegold. 1994. Chapter 9 in *Bailey and Scott's Diagnostic Microbiology*, 9th Ed. Mosby-Yearbook, Inc., St. Louis, MO.

Koneman, Elmer W., Stephen D. Allen, William M. Janda, Paul C. Schreckenberger, and Washington C. Winn, Jr. 1997. Chapter 14 in *Color Atlas and Textbook of Diagnostic Microbiology*, 5th Ed. J. B. Lippincott Company, Philadelphia, PA.

■ OBSERVATIONS AND INTERPRETATIONS

Record your observations and interpretations in the table below.

OBSERVATIONS AND INTERPRETATIONS

ORGANISM	REGION OF GROWTH	AEROTOLERANCE CATEGORY
Uninoculated Control		

Anaerobic Culture Methods

The following two exercises examine methods of cultivating organisms for which free oxygen must be reduced or eliminated. **Fluid Thioglycollate Medium** is a liquid medium designed to chemically reduce free oxygen and subsequently inhibit its diffusion from the atmosphere, thus providing an excellent environment for both anaerobes and microaerophiles. The **Anaerobic Jar** is a device that, when purged of its oxygen, creates an anaerobic environment, within which, any appropriate media (usually plates) can be incubated. Both systems create isolated inner anaerobic environments and can be incubated aerobically.

Although these systems are designed for cultivation of anaerobes and microaerophiles, they can be useful for observing specific aerotolerance categories. In these exercises you will be using both the anaerobic jar and thioglycolate broth to compare and categorize four organisms based on their oxygen tolerance properties. In Exercise 2-9 duplicate plates will be incubated aerobically for comparative purposes.

2-8 FLUID THIOGLYCOLLATE MEDIUM

Photographic Atlas Reference
Anaerobic Culture Methods Page 10

MATERIALS NEEDED FOR THIS EXERCISE

Per Student Group

- Five Fluid Thioglycollate Medium tubes
- Fresh cultures of:
  - *Pseudomonas aeruginosa*
  - *Clostridium sporogenes*
  - *Staphylococcus aureus*

PROCEDURE

Lab One

1. Obtain four Fluid Thioglycollate tubes and label them with your name, the date, medium, and organism.
2. Using your loop inoculate three broths with the provided organisms. Do not inoculate the fourth tube; it will be your control.
3. Incubate the tubes at 37°C for 24 to 48 hours.

Lab Two

1. Check the control tube for growth to assure sterility of the medium. Note any changes that may have occurred as a result of incubation, especially in the colored region at the surface.
2. Using the control as a comparison examine and note the location of the growth in all tubes.
3. Enter your observations and interpretations in the table provided.

REFERENCES

Allen, Stephen D., Jean A. Siders, and Linda M. Marler. 1985. Chapter 37 in *Manual of Clinical Microbiology*, 4th Ed. Edited by Edwin H. Lennette, Albert Balows, William J. Hausler, Jr., and H. Jean Shadomy. American Society for Microbiology, Washington, DC.

Forbes, Betty A., Daniel F. Sahm, and Alice S. Weissfeld. 2002. Pages 138–140 in *Bailey and Scott's Diagnostic Microbiology*, 11th Ed. Mosby-Yearbook, St. Louis, MO.

DIFCO Laboratories. 1984. Page 951 in *DIFCO Manual*, 10th Ed. DIFCO Laboratories, Detroit, MI.

Koneman, Elmer W., Stephen D. Allen, William M. Janda, Paul C. Schreckenberger, and Washington C. Winn, Jr. 1997. Chapter 14 in *Color Atlas and Textbook of Diagnostic Microbiology*, 5th Ed. J. B. Lippincott Company, Philadelphia, PA.

Power, David A. and Peggy J. McCuen. 1988. Page 261 and 311 in *Manual of BBL® Products and Laboratory Procedures*, 6th Ed. Becton Dickinson Microbiology Systems, Cockeysville, MD.

OBSERVATIONS AND INTERPRETATIONS

Record your observations and interpretations in the table below.

OBSERVATIONS AND INTERPRETATIONS

ORGANISM	LOCATION OF GROWTH IN MEDIUM	AEROTOLERANCE CATEGORY
Uninoculated Control		

2-9 ANAEROBIC JAR

Photographic Atlas Reference
Anaerobic Culture Methods Page 10

MATERIALS NEEDED FOR THIS EXERCISE

Per Class

- One anaerobic jar with gas generator packet (Becton Dickinson Microbiology Systems, Sparks, MD)

Per Group

- Two nutrient agar plates
- Fresh cultures of:
  - *Pseudomonas aeruginosa*
  - *Clostridium sporogenes*
  - *Staphylococcus aureus*

PROCEDURE

Lab One

1. Obtain two Nutrient Agar plates. Using a marking pen divide the bottom of each plate into three sectors.
2. Label each plate with your name, the date, and organism by sector.
3. Using your loop, inoculate the sectors of both plates with the organisms provided. (Inoculate with single streaks about one centimeter long.) Tape the lids on.
4. Place one plate in the anaerobic jar in an inverted position.
5. When all groups have placed their plates in the jar, discharge the packet as follows (or follow the instructions for your system).
   a. Stick the methylene blue strip on the wall of the jar.
   b. Open the packet and add 10 mL of distilled water.
   c. Place the open packet in the jar with the label facing inward.
   d. Immediately close the jar.
6. Place the second plate and the anaerobic jar in the 37°C incubator for 24 to 48 hours.

Lab Two

1. Examine and compare the growth on the plates. (Density of growth will be the most useful basis of comparison.)
2. Record your results and interpretations in the table provided.

REFERENCES

DIFCO Laboratories. 1984. Page 951 in *DIFCO Manual*, 10th Ed. DIFCO Laboratories, Detroit, MI.

Forbes, Betty A., Daniel F. Sahm, and Alice S. Weissfeld. 2002. Pages 516–517 in *Bailey and Scott's Diagnostic Microbiology*, 11th Ed. Mosby-Yearbook, St. Louis, MO.

Koneman, Elmer W., Stephen D. Allen, William M. Janda, Paul C. Schreckenberger, and Washington C. Winn, Jr. 1997. Chapter 14 in *Color Atlas and Textbook of Diagnostic Microbiology*, 5th Ed. J. B. Lippincott Company, Philadelphia, PA.

Power, David A. and Peggy J. McCuen. 1988. Page 261 and 311 in *Manual of BBL® Products and Laboratory Procedures*, 6th Ed. Becton Dickinson Microbiology Systems, Cockeysville, MD.

OBSERVATIONS AND INTERPRETATIONS

Using 0 for no growth and +, ++, and +++ for poor to good growth respectively, record your results in the table below.

OBSERVATIONS AND INTERPRETATIONS			
ORGANISM	GROWTH ON AEROBIC PLATE	GROWTH ON ANAEROBIC PLATE	AEROTOLERANCE CATEGORY

2-10 EFFECT OF TEMPERATURE ON MICROBIAL GROWTH

Bacteria have been discovered in habitats ranging from –10 degrees Celsius to more than 110 degrees Celsius. However, no single species occupies more than a small portion of this vast range. As such, each species is characterized by a minimum, maximum, and optimum temperature—collectively known as its **cardinal temperatures** (Figure 2-3). Minimum and maximum temperatures are, simply, the temperatures below and above which the organism will not survive. Optimum temperature is the temperature at which an organism shows the greatest growth over time—its highest growth rate.

Organisms that grow only below 20 degrees Celsius are called **psychrophiles.** Psychrophiles are very common in ocean, Arctic, and Antarctic habitats where the temperature remains permanently cold with little or no fluctuation. Organisms adapted to cold habitats that fluctuate from about 0 degrees to above 30 degrees Celsius are called **psychrotrophs.** Bacteria adapted to temperatures between 15 degrees and 45 degrees Celsius are known as **mesophiles.** Most bacterial residents in the human body as well as numerous human pathogens are mesophiles. **Thermophiles** are organisms adapted to temperatures above 40 degrees Celsius. They are typically found in composting organic material and in hot springs. Heat-adapted organisms that will not grow at temperatures below 40 degrees are called **obligate thermophiles;** those that will grow below 40 degrees are known as **facultative thermophiles.** Bacteria isolated from hot ocean floor ridges living between 65 and 110 degrees Celsius are called **extreme thermophiles.** Extreme thermophiles grow best above 80 degrees Celsius. Figure 2-4 illustrates bacterial temperature ranges and classifications.

In this exercise you will examine the growth characteristics of three organisms at five different temperatures. In addition, you will observe the influence of temperature on pigment production.

MATERIALS NEEDED FOR THIS EXERCISE

Per Class

- Five temperature sources (refrigerator/ice bath, incubator, hot water bath) set at 5°C, 20°C, 35°C, 50°C, and 65°C.
- Spectrophotometers (optional)
- Cuvettes (optional)

Per Student Group

- Twenty sterile nutrient broths
- Two trypticase soy agar (TSA) plates
- Three sterile transfer pipettes
- Three nonsterile transfer pipettes
- Fresh nutrient broth cultures of:
  - *Escherichia coli*
  - *Serratia marcescens*
  - *Bacillus stearothermophilus*

PROCEDURE

Lab One

1. Obtain twenty nutrient broths—one broth for each organism at each temperature plus one control for each temperature. Label them accordingly. Also obtain two TSA plates and label them 20°C and 35°C respectively.
2. Mix the cultures thoroughly and using a sterile pipette transfer a *single drop* of each to its appropriate tubes. (Since you will be comparing amount of growth at each

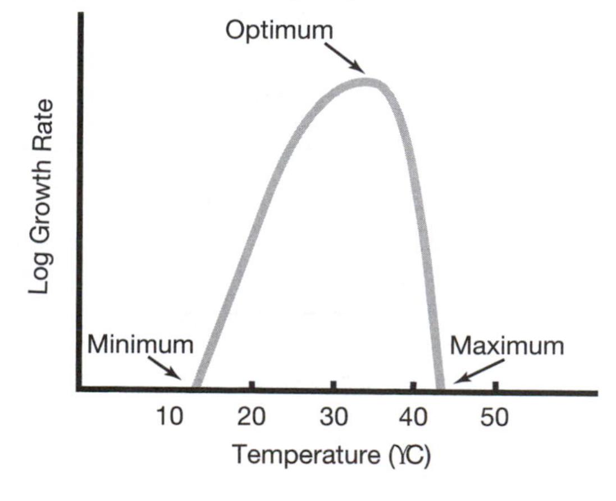

FIGURE 2-3 Typical Growth Range of a Mesophile
The "minimum" and "maximum" are temperatures beyond which no growth takes place. The "optimum" is the temperature at which growth rate is highest. After Prescott (1999).

FIGURE 2-4 Thermal Classifications of Bacteria
After Prescott (1999).

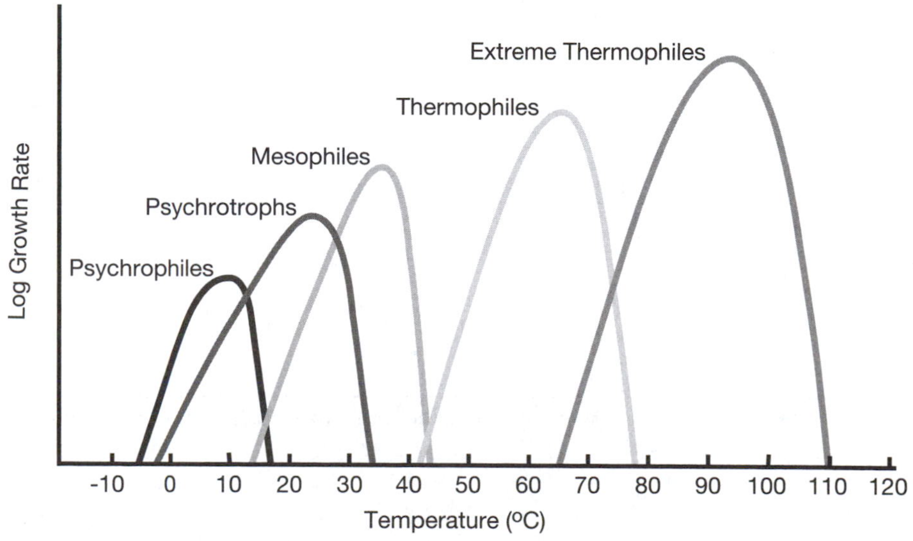

temperature, it is important to begin by transferring the same volume of culture to each broth.) Use the same pipette for all transfers with a single organism.

3. Using a simple zigzag pattern (as in Figure 2-1) inoculate each plate with *Serratia marcescens*.
4. Incubate all tubes in their appropriate temperatures for 24 to 48 hours. Incubate the plates in the 20°C and 35°C incubators respectively.

Lab Two (Without a spectrophotometer)

1. Clean the outside of all tubes with a tissue and place them in a test tube rack organized into groups by *organism*.
2. Gently shake each broth until uniform turbidity is achieved.
3. Examine the controls for turbidity. If no turbidity is present, place them in the rack with the test broths.
4. Compare all tubes in a group to each other and to a control. Rate each one as: 0, +, ++, or +++ according to its turbidity (0 is clear and +++ is very turbid). Record these in the table below.
5. Examine the plates incubated at different temperatures, compare the growth characteristics, and enter your results in the table provided on page 29.
6. Using the data from the table and the terms from the discussion on page 27, determine the cardinal temperatures and classification of each of the three test organisms.
7. **Optional:** On a separate piece of graph paper, plot the data ("+" values vs. temperature) of the three organisms.

Lab Two (Using a spectrophotometer)

1. Completely dry all tubes and place them in a test tube rack organized into groups by *organism*.
2. Gently shake each broth until uniform turbidity is achieved.
3. Set the spectrophotometer wavelength at 650 nm. If the spectrophotometer is digital set it to "absorbance."
4. Blank the spectrophotometer with the control for the first temperature group. Take the absorbance readings of all tubes in the group and enter the results in the table provided. Turbidity (a qualitative measure) increases with higher cell density. Measuring absorbance—which increases proportionately with turbidity—allows us to quantify the differences in growth. (Note: Your instructor may provide you with cuvettes for this portion of the exercise. Use a nonsterile transfer pipette to fill the cuvettes. After use, dispose of the cuvettes and transfer pipettes in an autoclave receptacle.)
5. Using the control for each group to blank the spectrophotometer, continue reading all of the broths and enter the results in the table provided below.
6. Examine the plates incubated at different temperatures, compare the growth characteristics, and enter your results in the table provided on page 29.
7. Using the data from the table and the terms from the discussion on page 27, determine the cardinal temperatures and classification of each of the three test organisms.
8. **Optional:** On a separate piece of graph paper, plot the data (absorbance vs. temperature) of the three organisms.

OBSERVATIONS AND INTERPRETATIONS

Record your absorbance values or qualitative scores for each organism at each temperature.

OBSERVATIONS AND INTERPRETATIONS						
	Broth Data					
Organism	5°C	20°C	35°C	50°C	65°C	Classification
Uninoculated Control						

Enter the absorbance if using a spectrophotometer. Enter visual readings as 0, +, ++, or +++.

Observations and Interpretations

Record the cultural characteristics of the *Serratia marcescens* incubated at two temperatures.

OBSERVATIONS AND INTERPRETATIONS	
Plate Data	
Incubation Temperature	**Description of Growth**
20°C	
35°C	

References

Forbes, Betty A., Daniel F. Sahm, and Alice S. Weissfeld. 2002. Page 142 in *Bailey and Scott's Diagnostic Microbiology*, 11th Ed. Mosby-Yearbook, St. Louis, MO.

Holt, John G. (Editor). 1994. *Bergey's Manual of Determinative Bacteriology*, 9th Ed. Williams and Wilkins, Baltimore, MD.

Koneman, Elmer W., Stephen D. Allen, William M. Janda, Paul C. Schreckenberger, and Washington C. Winn, Jr. 1997. *Color Atlas and Textbook of Diagnostic Microbiology*, 5th Ed. J. B. Lippincott Company, Philadelphia, PA.

Moat, Albert G., John W. Foster, and Michael P. Spector. 2002. Pages 597–601 in *Microbial Physiology*, 4th Ed. Wiley-Liss, Inc. New York, NY.

Prescott, Lansing M., John P. Harley, and Donald A. Klein. 2005. Chapter 6 in *Microbiology*, 6th Ed., WCB McGraw-Hill, Boston, MA.

Varnam, Alan H. and Malcolm G. Evans. 2000. *Environmental Microbiology*. ASM Press, Washington, DC.

White, David. 2000. Pages 384–387 in *The Physiology and Biochemistry of Prokaryotes*, 2nd Ed. Oxford University Press, New York, NY.

2-11 EFFECT OF pH ON MICROBIAL GROWTH

pH is the conventional means of expressing the concentration of hydrogen ions in a solution. pH values range from 0 to 14 and are actually negative logarithms of the hydrogen ion concentration in moles per liter. Mathematically the formula appears as:

$$pH = -\log [H^+].$$

Pure water contains 10^{-7} moles of hydrogen ions per liter and has a pH of 7. As hydrogen ions increase the solution becomes more acidic and the pH goes down (Table 2-1).

Bacteria live in habitats throughout the pH spectrum; however, the range of most individual species is small. Like temperature and salinity, pH tolerance is used as a means of classification. The three major classifications are **acidophiles, neutrophiles,** and **alkaliphiles**. Acidophiles are organisms adapted to grow well in environments below about pH 5.5. Neutrophiles are organisms that prefer pH levels between 5.5 and 8.5, and alkaliphiles live above pH 8.5.

Under normal circumstances bacteria maintain a near-neutral internal environment regardless of their habitat. pH changes outside an organism's range may destroy necessary membrane potential (in the production of ATP) and damage vital enzymes beyond repair. This **denaturing** of cellular enzymes involves conformational changes in the proteins' tertiary structure and is usually lethal to the cell.

In this exercise you will cultivate and classify three organisms based on their minimum, maximum, and optimum growth pH.

Materials Needed For This Exercise

Per Student Group

- pH adjusted Nutrient Broths as follows: pH 2, pH 4, pH 6, pH 8, and pH 10
- Spectrophotometer (optional)
- Cuvettes (optional)
- Three sterile transfer pipettes
- Three nonsterile transfer pipettes
- Fresh nutrient broth cultures of:
  - *Lactobacillus plantarum*
  - *Lactococcus lactis*
  - *Staphylococcus aureus*

Procedure

Lab One

1. Obtain four tubes of each pH broth (20 tubes total)—one of each pH per organism plus one of each for controls. Label them accordingly.
2. Mix the cultures thoroughly and using a sterile pipette transfer a *single drop* of each to its appropriate tubes. (Since you will be comparing amount of growth at each pH, it is important to begin by transferring the same volume of culture to each broth.) Use the same pipette for all transfers with a single organism.
3. Incubate all tubes at 37°C for 48 hours.

TABLE 2-1 pH SCALE

Acidity/ Alkalinity	pH	H+ Concentration in Moles/Liter	Common Examples	Bacterial Ranges
↑	0	10^{0}	Nitric acid	
	1	10^{-1}	Stomach acid	↑
	2	10^{-2}	Lemon juice	
	3	10^{-3}	Vinegar, cola	
	4	10^{-4}	Tomatoes, orange juice	Acidophiles
	5	10^{-5}	Black coffee	
Acidic	6	10^{-6}	Urine	Neutrophiles
Neutral	7	10^{-7}	Pure water	
Alkaline	8	10^{-8}	Sea water	Alkaliphiles
	9	10^{-9}	Baking soda	
	10	10^{-10}	Soap, milk of magnesia	
	11	10^{-11}	Ammonia	
	12	10^{-12}	Lime water [$Ca(OH)_2$]	
↓	13	10^{-13}	Household bleach	↓
	14	10^{-14}	Drain cleaner	

Lab Two (Without a spectrophotometer)

1. Clean the outside of all tubes with a tissue and place them in a test tube rack organized into groups by *organism*.
2. Gently shake each broth until uniform turbidity is achieved.
3. Compare all tubes in a group to each other and to a control. Rate each one as: 0, +, ++, or +++ according to its turbidity (0 is clear and +++ is very turbid).
4. Using the data from Table 2-1 and the terms from the discussion on page 30, determine the range and classification of each of the three test organisms.
5. **Optional**: On a separate sheet of graph paper, plot the data ("+" values vs. pH) of the three organisms.

Lab Two (Using a spectrophotometer)

1. Clean the outside of all tubes with a tissue and place them in a test tube rack organized into groups by *organism*.
2. Gently shake each broth until uniform turbidity is achieved. Continue to agitate the tubes as needed to keep solids from settling to the bottom.
3. Set the spectrophotometer wavelength at 650 nm. If the spectrophotometer is digital set it to "absorbance."
4. Blank the spectrophotometer with the control for the first pH group. Take the absorbance readings of all the tubes in the group and enter the results in the table provided. Turbidity (a qualitative measure) increases with higher cell density. Measuring absorbance—which increases proportionately with turbidity—allows us to quantify the differences in growth. (**Note:** Your instructor may provide you with cuvettes for this portion of the exercise. Use a nonsterile transfer pipette to fill the cuvettes. After use, dispose of the cuvettes and transfer pipettes in an autoclave receptacle.)
5. Using the control for each group to blank the spectrophotometer, continue reading all of the broths and enter the results in the table provided.
6. Using the data from Table 2-1 and the terms from the discussion on page 30, determine the range and classification of each of the three test organisms.
7. **Optional**: On a separate sheet of graph paper, plot the data (absorbance vs. pH) of the three organisms.

REFERENCES

Forbes, Betty A., Daniel F. Sahm, and Alice S. Weissfeld. 2002. Page 142–143 in *Bailey and Scott's Diagnostic Microbiology*, 11th Ed. Mosby-Yearbook, St. Louis, MO.

Holt, John G. (Editor). 1994. *Bergey's Manual of Determinative Bacteriology*, 9th Ed. Williams and Wilkins, Baltimore, MD.

Koneman, Elmer W., Stephen D. Allen, William M. Janda, Paul C. Schreckenberger, and Washington C. Winn, Jr. 1997. *Color Atlas and Textbook of Diagnostic Microbiology*, 5th Ed. J. B. Lippincott Company, Philadelphia, PA.

Moat, Albert G., John W. Foster, and Michael P. Spector. 2002. Pages 382–383 in *Microbial Physiology*, 4th Ed. Wiley-Liss, Inc. New York, NY.

Varnam, Alan H. and Malcolm G. Evans. 2000. *Environmental Microbiology*. ASM Press, Washington, DC.

White, David. 2000. Pages 384–387 in *The Physiology and Biochemistry of Prokaryotes*, 2nd Ed. Oxford University Press, New York, NY.

OBSERVATIONS AND INTERPRETATIONS

Record your absorbance values or qualitative scores for each organism at each pH. Classify each organism using the terms from the discussion above.

OBSERVATIONS AND INTERPRETATIONS

ORGANISM	pH 2	pH 4	pH 6	pH 8	pH 10	CLASSIFICATION
Uninoculated Control						

Enter the absorbance if using a spectrophotometer. Enter visual readings as 0, +, ++, or +++.

2-12 EFFECT OF OSMOTIC PRESSURE ON MICROBIAL GROWTH

Water is essential to all forms of life. It is not only the principal component of cytoplasm, but also an essential source of electrons and hydrogen ions. Microorganisms, like plants, require water to maintain cellular **turgor pressure**. Whereas eukaryotic animal cells burst with a constant influx of water, prokaryotes require it to prevent shrinking of the cell resulting in separation of the membrane from the cell wall (a condition called **plasmolysis**).

Many bacteria regulate turgor pressure by transporting in and maintaining a relatively high cytoplasmic potassium or sodium ion concentration, thus creating a concentration gradient that promotes inward diffusion of water. **Compatible solutes** composed primarily of amino acids are also transported into the cell to help maintain turgor pressure and provide essential building blocks for cellular components. For bacteria living in saline habitats the job of maintaining turgor pressure is continuous.

Irrespective of a cell's efforts to control its internal environment, natural forces will cause water to move through the semipermeable membrane from an area of low **solute** concentration to an area of high solute concentration. In a solution where solute concentration is low, water concentration is high and *vice versa*. Therefore, water moves from where its concentration is high to where its concentration is low. This process is called **osmosis.** If a bacterial cell is placed into a solution that is **hyposmotic,** there will be net movement of water down its gradient and into the cell. If an organism is placed into a **hyperosmotic** solution, there will be a net diffusion of water out of the cell. For bacteria living in an **isosmotic** solution, water will tend to move in both directions equally, that is, there is no net movement (Figure 2-5).

Bacteria constitute a diverse group of organisms and, as such, have developed many adaptations for survival.

Microorganisms tend to have a distinct range of salinities that is optimal for growth with little or no survival outside that range. For example, some bacteria called **halophiles** grow optimally in NaCl concentrations 3% or higher. **Extreme halophiles** are organisms with specialized cell membranes and enzymes that require salt concentrations from 15% up to about 25% and will not survive where salinity is lower. Except for a few **osmotolerant** bacteria, which will grow over a wide range of salinities, most bacteria live where NaCl concentrations are less than 3%.

In this exercise you will grow two human commensal bacteria at a variety of NaCl concentrations to determine the maximum tolerance for each organism. *Staphylococcus aureus* is an inhabitant of human skin and nasal passages. *Escherichia coli* is a common bacterial species living in human intestines. See if you can predict a correlation between the osmotolerance of each organism and the areas of the human body where it is most often found.

MATERIALS NEEDED FOR THIS EXERCISE

Per Student Group

- Nutrient broths containing 1%, 3%, 5%, 7%, 9%, and 11% NaCl respectively
- Spectrophotometer (optional)
- Cuvettes (optional)
- Sterile transfer pipettes
- Fresh broth cultures of:
  Staphylococcus aureus
  Escherichia coli

PROCEDURE

Lab One

1. Obtain three tubes of each NaCl broth—one of each concentration per organism plus one of each for controls—and label them accordingly.
2. Mix the cultures thoroughly and using a sterile pipette transfer a *single drop* of each to its appropriate tubes. (Since you will be comparing amount of growth at each NaCl concentration, it is important to begin by transferring the same volume of culture to each broth.) Use the same pipette for all transfers with a single organism.
3. Incubate all tubes at 37°C for 48 hours.

■ **FIGURE 2-5 The Effect of Tonicity on Bacterial Cells**
In a hypotonic environment the net movement of water (arrows) will be into the cell. In an isotonic environment there is no net movement. (Actually the water is moving equally in both directions.) In a hypertonic environment the net movement is outward and results in plasmolysis. Note the shrinking membrane (CM) and the rigid cell wall (CW) in the hypertonic solution.

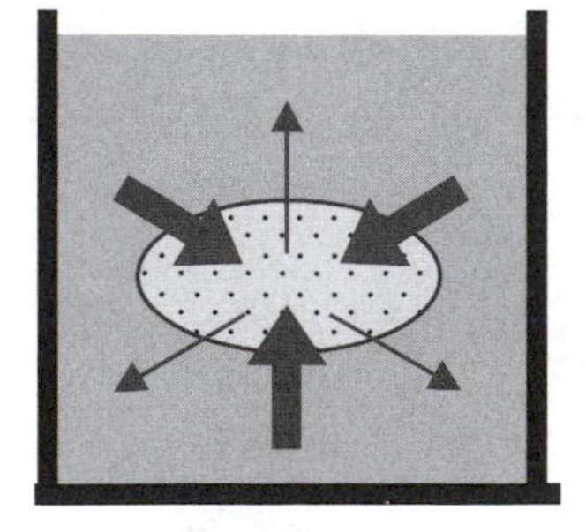

Hypotonic Environment

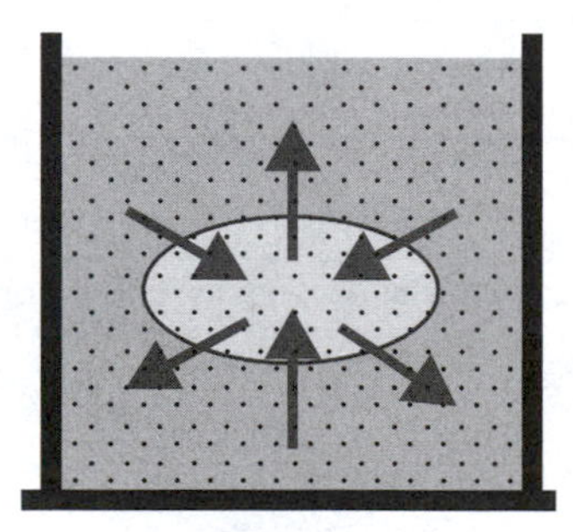

Isotonic Environment

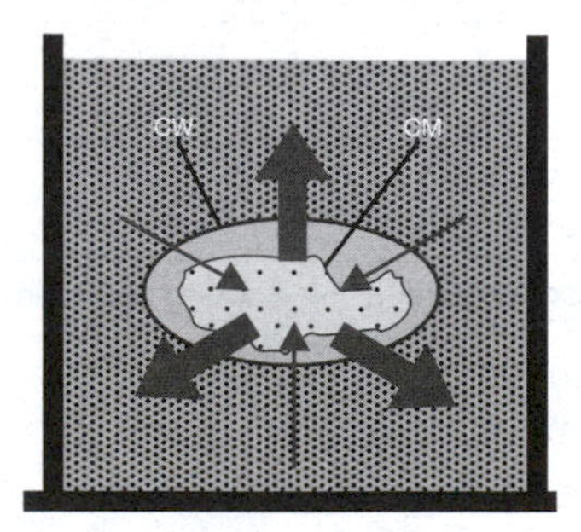

Hypertonic Environment

Lab Two (Without a spectrophotometer)

1. Clean the outside of all tubes with a tissue and place them in a test tube rack organized into groups by *organism*.
2. Gently shake each broth until uniform turbidity is achieved.
3. Compare all tubes in a group to each other and to the control tube. Rate each one as: 0, +, ++, or +++ according to its turbidity (0 is clear and +++ is very turbid).
4. Enter your results in the table provided.
5. **Optional:** On a separate sheet of graph paper, plot the "+" values vs. % NaCl of both organisms.

Lab Two (Using a spectrophotometer)

1. Clean the outside of all tubes with a tissue and place them in a test tube rack organized into groups by *organism*.
2. Gently shake each broth until uniform turbidity is achieved.
3. Set the spectrophotometer wavelength at 650 nm. If the spectrophotometer is digital set it to "absorbance."
4. Blank the spectrophotometer with the uninoculated control for the first group of tubes. Take the absorbance readings of all the tubes in the group and enter the results in the table below. Turbidity (a qualitative measure) increases with higher cell density. Measuring absorbance—which increases proportionately with turbidity—allows us to quantify the differences in growth. (Note: Your instructor may provide you with cuvettes for this portion of the exercise. Use a nonsterile transfer pipette to fill the cuvettes. After use, dispose of the cuvettes and transfer pipettes in an autoclave receptacle.)
5. Using the control for each group to blank the spectrophotometer, continue reading all of the broths and enter the results in the table provided.
6. **Optional:** On a separate piece of graph paper, plot the absorbance vs. % NaCl of both organisms.

References

Forbes, Betty A., Daniel F. Sahm, Alice S. Weissfeld. 1998. *Bailey & Scott's Diagnostic Microbiology*, 10th Ed. Mosby, Inc., St. Louis, MO.

Holt, John G. (Editor). 1994. *Bergey's Manual of Determinative Bacteriology*, 9th Ed. Williams and Wilkins, Baltimore, MD.

Koneman, Elmer W., Stephen D. Allen, William M. Janda, Paul C. Schreckenberger, and Washington C. Winn, Jr. 1997. *Color Atlas and Textbook of Diagnostic Microbiology*, 5th Ed. J. B. Lippincott Company, Philadelphia, PA.

Moat, Albert G., John W. Foster, and Michael P. Spector. 2002. Pages 582–587 in *Microbial Physiology*, 4th Ed. Wiley-Liss, Inc. New York, NY.

Varnam, Alan H. and Malcolm G. Evans. 2000. *Environmental Microbiology*. ASM Press, Washington, DC.

White, David. 2000. Pages 388–394 in *The Physiology and Biochemistry of Prokaryotes*, 2nd Ed. Oxford University Press, New York, NY.

Observations and Interpretations

Record your absorbance values or qualitative scores for each organism at each NaCl concentration.

OBSERVATIONS AND INTERPRETATIONS

Organism	NaCl Concentration					
	1%	3%	5%	7%	9%	11%

Enter the absorbance if using a spectrophotometer. Enter visual readings as 0, +, ++, or +++.

Control of Microorganisms

As demonstrated in earlier exercises knowledge of factors that affect microbial growth are of great importance to microbiologists. Of nearly equal importance to microbiologists is the ability to reliably stop or inhibit unwanted growth. The following three exercises will examine microbial control methods and related aspects.

Most microbial control agents function by damaging cell membranes or proteins, thus disrupting cellular transport and enzyme activity. The effectiveness of most control systems is measured against microbial ability to produce **spores**. Killing most **vegetative cells** is not particularly difficult, but killing spores usually takes special techniques and equipment.

There are many commonly used control methods ranging from simple scrubbing to complete destruction of all living matter. The choice of methods or agents depends on the level of control needed and the circumstances under which the work is being done. In this unit you will examine the effects of radiation and a variety of antimicrobial agents including antibiotics, antiseptics, and disinfectants.

2-13 THE LETHAL EFFECT OF ULTRAVIOLET LIGHT ON MICROBIAL GROWTH

Ultraviolet (UV) light is a type of **electromagnetic energy**. Like all electromagnetic energy, UV travels in waves and is distinguishable from all others by its **wavelength**. Wavelength is the distance between adjacent wave crests.

Ultraviolet light is divided into three groups categorized by wavelength—UV-A, UV-B, and UV-C. UV-A wavelengths are the longest and range from 315 to 400 nm. UV-B wavelengths are between 280 and 315 nm. UV-C wavelengths, the most detrimental to bacteria, range from 100 to 280 nm. Bacterial exposure to UV-C for more than a few minutes usually results in death of the organism from irreparable DNA damage. For a discussion on the mutagenic effects of UV and DNA repair, refer to Exercise 8-2.

This exercise examines the effect of UV exposure on two *Bacillus subtilis* cultures (24-hour and 7-day) and one 24-hour *Escherichia coli* culture. Due to the large number of plates to be treated it will be necessary to divide the work among six groups of students. Refer to Table 2-2 for assignments.

MATERIALS NEEDED FOR THIS EXERCISE

Per Student Group

- An ultraviolet lamp with appropriate shielding
- Cardboard strips or disks to cover plates
- Three Tryptic Soy Agar (TSA) plates (six for group one)
- Stopwatch or electronic timer
- Sterile Nutrient Broths (three per group; six for group one)
- Sterile cotton swabs (three per group; six for group one)
- Slant cultures of:
  - *Bacillus subtilis* (24 hour culture)
  - *Bacillus subtilis (7 day culture)*
  - *Escherichia coli (24 hour culture)*

PROCEDURE

Lab One

1. Enter your group number and exposure time here. ________________(see Table 2-2).
2. Obtain three TSA plates and label the bottom of each with the name of the organism to be inoculated and your group number. Draw a line to divide the plates in half and label the sides "A" and "B".

TABLE 2-2 GROUP ASSIGNMENTS BY NUMBER

ORGANISM	NO UV	5 MIN.	10 MIN.	15 MIN.	20 MIN.	25 MIN.	30 MIN.
Bacillus subtilis (24-hr culture)	1	1	2	3	4	5	6
Bacillus subtilis (7-day culture)	1	1	2	3	4	5	6
Escherichia coli (24-hr culture)	1	1	2	3	4	5	6

3. Pour a tube of Nutrient Broth into each slant and mix gently. Be careful to do this aseptically and to not overflow the tube.
4. Dip a sterile cotton swab into the broth of one culture and wipe the excess on the inside of the tube. Inoculate the appropriate plate by spreading the organism over the entire surface of the agar. Do this by streaking the plate surface completely three times, rotating it one-third turn between streaks. When incubated this will form a bacterial lawn.
5. Repeat step 4 with the other two organisms and plates.
6. Place a paper towel on the table next to the UV lamp and soak it with disinfectant.
7. Place your plates under the UV lamp and set the covers, open side down, on the disinfectant-soaked towel. Cover the "B" half of the plates with the cardboard as shown in Figure 2-6.
8. Turn the lamp on for the prescribed time. **CAUTION! BE SURE THE PROTECTIVE SHIELD IS IN PLACE AND DO NOT LOOK AT THE LIGHT WHILE IT IS ON.** Immediately replace the plate covers.
9. Invert and incubate the plates at 37°C for 24 to 48 hours.

Lab Two

1. Remove your plates from the incubator and observe for growth. Side "B" should be covered with a bacterial lawn. In the table below, under your exposure time, record the growth on side "A" of each plate. Enter "0" if you observe no growth, "+" for poor growth, "++" for moderate growth, and "+++" for abundant growth.
2. Your instructor may provide an overhead transparency or blackboard space for you to enter your data.
3. Complete the table using the data provided by the rest of the class.
4. **Optional:** On a separate sheet of graph paper, construct a graph representing growth vs. UV exposure time for all three organisms.

■ FIGURE 2-6 Plates Shielded for UV Exposure
Place the three plates under the UV lamp with the covers removed and the cardboard shield covering half of each plate as shown. Make sure the Petri dish covers are placed open side down on a disinfectant soaked towel.

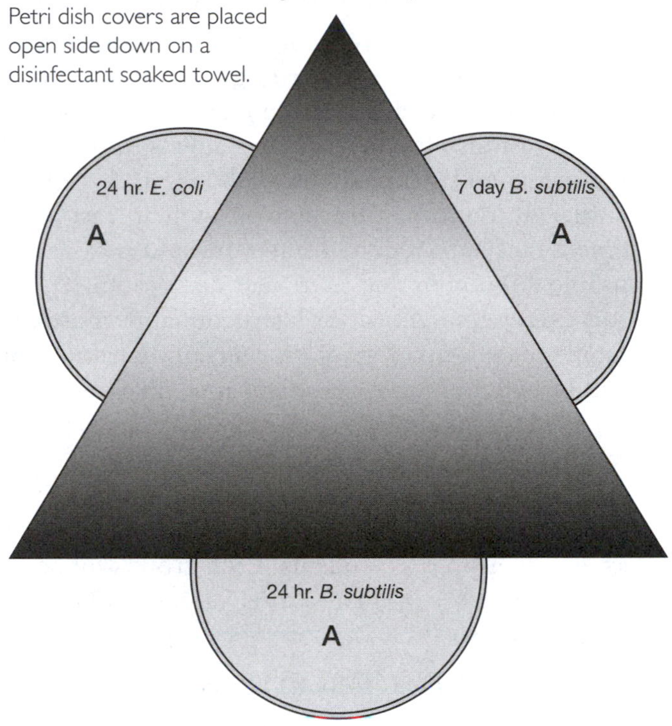

REFERENCES

Lewin, Benjamin. 2003. *Genes VIII*. Prentice Hall, Upper Saddle River, NJ.

Varnam, Alan, H. and Malcolm G. Evans. 2000. *Environmental Microbiology*. ASM Press, Washington, DC.

■ OBSERVATIONS AND INTERPRETATIONS

Enter your class data in the table below. Score the relative amount of growth on each plate. Use the symbols "0" for no growth, "+++" for abundant growth, and "+" and "++" for degrees of growth in between.

OBSERVATIONS AND INTERPRETATIONS							
ORGANISM	NO UV	5 MIN.	10 MIN.	15 MIN.	20 MIN.	25 MIN.	30 MIN.

2-14 USE-DILUTION: MEASURING DISINFECTANT EFFECTIVENESS

Use-dilution is a standard procedure used to measure disinfectant effectiveness against microorganisms. In this procedure, organisms are exposed to varying concentrations (dilutions) of test disinfectants for ten minutes, transferred to growth media and incubated for 48 hours. If the solution is sufficient to prevent microbial growth at least 95% of the time it meets the required standards and is considered a usable dilution of that particular disinfectant.

In this exercise, modified for instructional purposes, we will examine the effectiveness of varying concentrations of household bleach, hydrogen peroxide, ethyl alcohol, and isopropyl alcohol on *Staphylococcus aureus*, *Bacillus cereus*, and *Mycobacterium smegmatis*. Because of the many combinations of disinfectants and cultures being attempted in this lab exercise it will be necessary to divide the tasks among groups of students. Each group will be responsible for one organism and three dilutions of one disinfectant. Refer to Table 2-3 for your assignments.

Material Needed For This Exercise

Per Student Group

- 100 mL flask of sterile deionized water
- Disinfectants: (Three concentrations of one disinfectant per group. See Table 2-3.)
  - 0.1%, 1.0%, and 10.0% household bleach
  - 1%, 2%, and 3% hydrogen peroxide
  - 25%, 50%, and 75% isopropyl alcohol
  - 25%, 50%, and 75% ethyl alcohol
- Eight sterile 60 mm Petri dishes
- One 60 mm Petri dish containing sterile ceramic beads (available from Key Scientific, Round Rock, Texas)
- One sterile 100 mm Petri dish containing filter paper
- Five sterile nutrient broth tubes
- Needle-nose forceps
- Small beaker with alcohol (for flaming forceps)
- Fresh broth cultures of (one per group):
  - *Staphylococcus aureus*
  - *Bacillus cereus*
  - *Mycobacterium smegmatis*

Procedure

In this exercise you will examine the effectiveness of varying concentrations of household bleach, hydrogen peroxide, ethyl alcohol, and isopropyl alcohol on *Staphylococcus aureus*, *Bacillus cereus*, and *Mycobacterium smegmatis*. Tasks are divided between 12 groups of students. Each group will be responsible for one organism and three dilutions of one disinfectant. Refer to Table 2-3 for your assignments.

Lab One

1. Enter your organism here ______________________.
2. Enter your disinfectant here ______________________.
3. Obtain a Petri dish of sterile ceramic beads, one broth culture, one plate containing sterile filter paper, eight 60 mm plates, the three dilutions of your assigned disinfectant, and five sterile nutrient broths. Prepare a small beaker of alcohol for flaming the forceps.
4. Properly label and place the materials in front of you in this order: culture, filter paper plate, three 60 mm plates containing the three disinfectant dilutions, five 60 mm plates containing deionized water (sufficient to cover the beads), and five sterile broths. Refer to the procedural diagram in Figure 2-7.
5. Gently mix the broth culture until uniform turbidity is achieved.
6. Alcohol-flame your forceps and place four ceramic beads into the broth culture. (Note: Always use proper aseptic technique and **sterilize the forceps before all transfers in this exercise.**)
7. After one minute decant the broth into a flask of disinfectant and dispense the beads onto the sterile filter paper. Decant as much of the broth as you can to avoid wetting the filter paper excessively. You may need to "coax" the beads out of the tube with a sterile inoculating loop. Spread the beads apart on the paper using sterile forceps and allow them sufficient time to dry.
8. When the beads are dry (about 5 minutes), place three beads into the three disinfectant dilutions in 60 mm plates. Mark the time here__________.
9. After 10 minutes remove the beads and place them into different 60 mm plates containing rinse water. It is important that all groups use the same exposure times. Allow them to soak for one minute.

■ Each group is responsible for one organism and three dilutions of one disinfectant.

TABLE 2-3 GROUP ASSIGNMENTS

Disinfectant	*Staphylococcus aureus*	*Bacillus cereus*	*Mycobacterium smegmatis*
Bleach	1	2	3
Hydrogen Peroxide	4	5	6
Isopropyl Alcohol	7	8	9
Ethyl Alcohol	10	11	12

■ **FIGURE 2-7 Procedural Diagram for Use-Dilution**

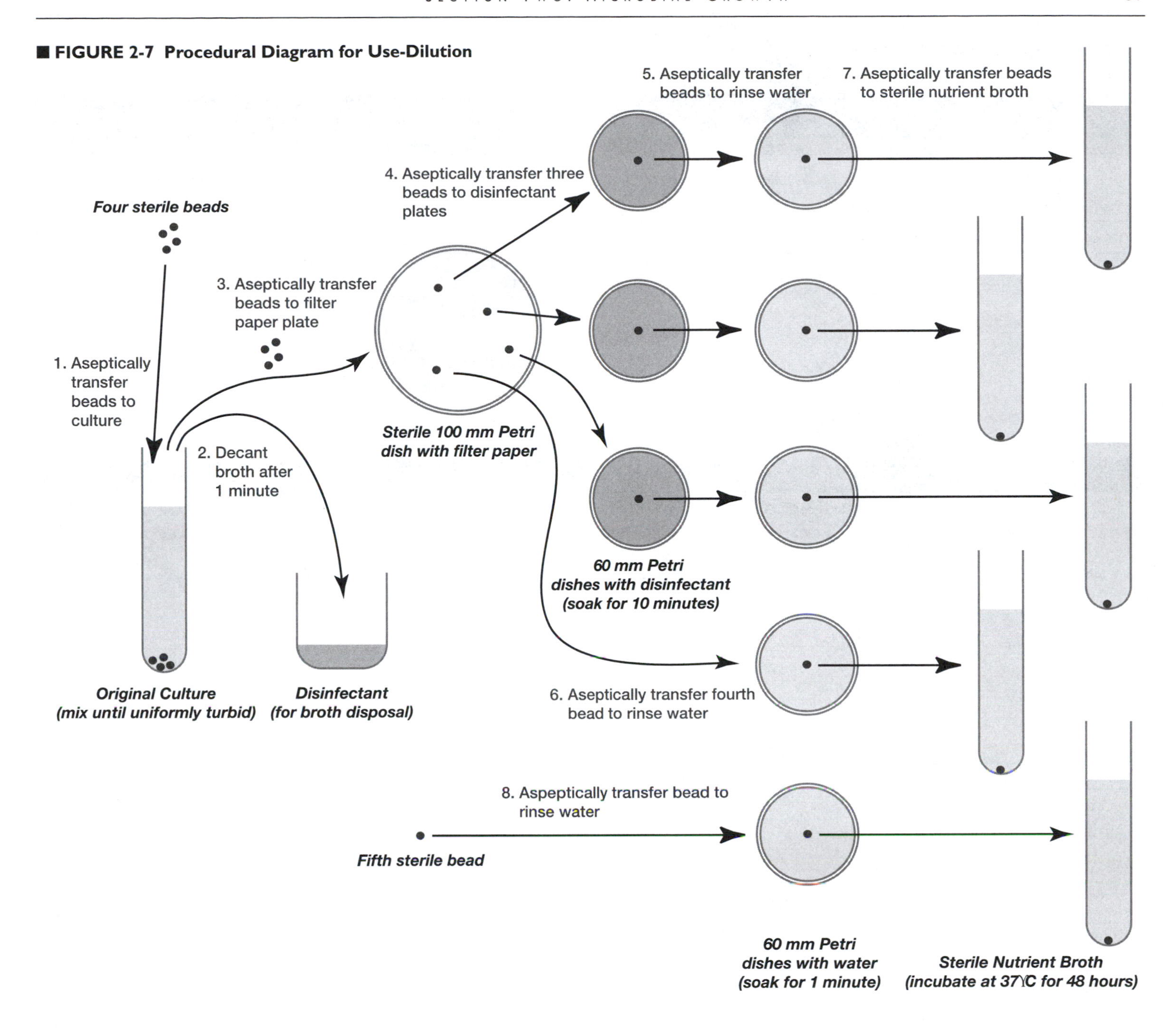

10. Aseptically remove the beads with the forceps and place them into the appropriately labeled Nutrient Broths.
11. Place the fourth bead containing culture into the fourth rinse plate. After one minute inoculate the fourth broth with it.
12. Place the fifth sterile bead (again using sterile technique) into the fifth sterile water rinse for one minute and then transfer it to the fifth sterile broth.
13. Incubate all broths at 37°C for 48 hours.

Lab Two

1. Remove all broths from the incubator and observe for any growth.
2. Closely examine your controls. In order to proceed, your inoculated bead must have produced growth and your uninoculated (sterile) bead must have produced no growth. If both of these requirements have been met, you may proceed.
3. Examine the remainder of your beads and using "G" to indicate growth and "NG" to indicate no growth, enter your results in the tables provided.

REFERENCE

Murray, Patrick R., *et al.* 1999. Page 138 in *Manual of Clinical Microbiology*, 7th Ed. American Society for Microbiology, Washington, DC.

Observations and Interpretations

Enter your individual data in the tables below.

OBSERVATIONS AND INTERPRETATIONS	
Controls	✓
Sterile bead (must have no growth)	
Inoculated bead (must have growth)	

OBSERVATIONS AND INTERPRETATIONS	
Disinfectant Solution	**Growth**

G = Growth, NG = No Growth

Observations and Interpretations

Enter class data in the table below.

OBSERVATIONS AND INTERPRETATIONS												
Organism	**Household Bleach**			**Hydrogen Peroxide**			**Ethyl Alcohol**			**Isopropyl Alcohol**		
	0.1%	**1%**	**10%**	**1%**	**2%**	**3%**	**25%**	**50%**	**75%**	**25%**	**50%**	**75%**

G = Growth, NG = No Growth

2-15 ANTIMICROBIAL SUSCEPTIBILITY TEST (KIRBY-BAUER METHOD)

Antimicrobial susceptibility testing is a standardized method used to measure the effectiveness of antibiotics and other chemotherapeutic agents on pathogenic microorganisms. In many cases, it is an essential tool in prescribing appropriate treatment. In this exercise you will test the susceptibility of *Escherichia coli* and *Staphylococcus aureus* strains to penicillin, streptomycin, tetracycline, and chloramphenicol.

Photographic Atlas Reference
Antibiotic Sensitivity Page 89

MATERIALS NEEDED FOR THIS EXERCISE

Per Student Pair

- Two Mueller-Hinton agar plates
- Streptomycin, tetracycline, penicillin and chloramphenicol antibiotic disks
- Antibiotic disk dispenser or forceps for placement of disks
- Small beaker of alcohol (for sterilizing forceps)
- Sterile cotton swabs
- One metric ruler
- Zone diameter interpretive table published by the National Committee for Clinical Laboratory Standards (NCCLS)
- Sterile saline (0.85%)
- Sterile Pasteur pipettes
- One McFarland 0.5 standard with card
- Fresh broth cultures of:
  - *Escherichia coli*
  - *Staphylococcus aureus*

PROCEDURE

Lab One

1. Gently mix a broth culture and the McFarland standard until they reach their maximum turbidity.
2. Holding the culture and McFarland standard upright in front of you, place the card behind them so that you can see the black line through the liquid in the tubes. As you can see, the line becomes distorted by the turbidity in the tubes (Figure 2-8). Use the black line to compare the turbidity level of the two tubes. Dilute the broth with the sterile saline until it appears to have the same level of turbidity as the standard.
3. Repeat this process with the other broth.
4. Dip a sterile swab into the *E. coli* broth and wipe off the excess on the inside of the tube.

■ FIGURE 2-8 McFarland Turbidity Standard Comparison
Comparison of a McFarland turbidity standard to three broths with varying degrees of turbidity. There are eleven McFarland standards (0.5 to 10), each of which contains a specific percentage of precipitated barium sulfate to produce turbidity. In the Kirby-Bauer procedure, the test culture is diluted to match the 0.5 McFarland standard (roughly equivalent to 1.5×10^8 cells per mL) before inoculating the plate. Comparison is made visually by placing a card containing sharp black lines behind the tubes.

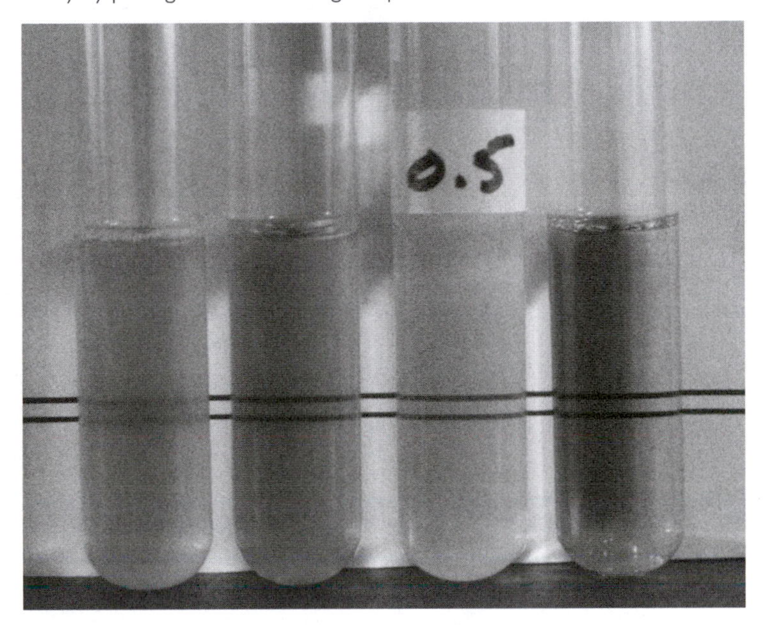

5. Inoculate a Mueller-Hinton plate with *E. coli* by streaking the entire surface of the agar three times with the swab. (**Note:** Rotate the plate 1/3 turn between streaks.)
6. Using a fresh sterile swab inoculate the other plate with *S. aureus*.
7. Label the plates with the organisms' names, your name, and the date.
8. Apply the streptomycin, tetracycline, penicillin and chloramphenicol discs to the agar surface of each plate. You can apply the disks either singly using sterile forceps, or with a dispenser. Be sure to space the disks sufficiently (4 to 5 cm) to prevent overlapping zones of inhibition.
9. Press each disc gently with sterile forceps so that it makes full contact with the agar surface.
10. Invert the plates and incubate them aerobically at 35°C for 16 to 18 hours. Have a volunteer in the group remove and refrigerate the plates at the appropriate time.

Lab Two

1. Remove the plates from the incubator and measure the diameter of the inhibition zones. It may be helpful to use a colony counter when taking measurements.
2. Using the Zone Diameter Interpretive Chart (available from NCCLS, Wayne, PA) or tables provided with your antibiotic discs, as a guide record your results in the table provided.

Observations and Interpretations

Record the zone diameters in mm. Enter "S" if the organism is susceptible to the antibiotic, "R" if it is resistant.

OBSERVATIONS AND INTERPRETATIONS								
Organism	**Streptomycin**		**Tetracycline**		**Penicillin**		**Chloramphenicol**	
	Zone Diameter	**S/R**	**Zone Diameter**	**S/R**	**Zone Diameter**	**S/R**	**Zone Diameter**	**S/R**

References

Collins, C. H., Patricia M. Lyne, J. M. Grange. 1995. Page 128 in *Collins and Lyne's Microbiological Methods*, 7th Ed. Butterworth-Heinemann, UK.

Ferraro, Mary Jane and James H. Jorgensen. 2003. Chapter 15 in *Manual of Clinical Microbiology*, 8th Ed., edited by Patrick R. Murray, Ellen Jo Baron, James H. Jorgensen, Michael A. Pfaller, and Robert H. Yolken, ASM Press, Washington, DC.

Forbes, Betty A., Daniel F. Sahm, and Alice S. Weissfeld. 2002. Pages 236–240 in *Bailey and Scott's Diagnostic Microbiology*, 11th Ed. Mosby-Yearbook, St. Louis, MO.

Koneman, Elmer W., Stephen D. Allen, William M. Janda, Paul C. Schreckenberger, and Washington C. Winn, Jr. 1997. Page 818–822 in *Color Atlas and Textbook of Diagnostic Microbiology*, 5th Ed. J. B. Lippincott Company, Philadelphia, PA.

Power, David A. and Peggy J. McCuen. 1988. Page 204 in *Manual of BBL® Products and Laboratory Procedures*, 6th Ed. Becton Dickinson Microbiology Systems, Cockeysville, MD.

SECTION THREE

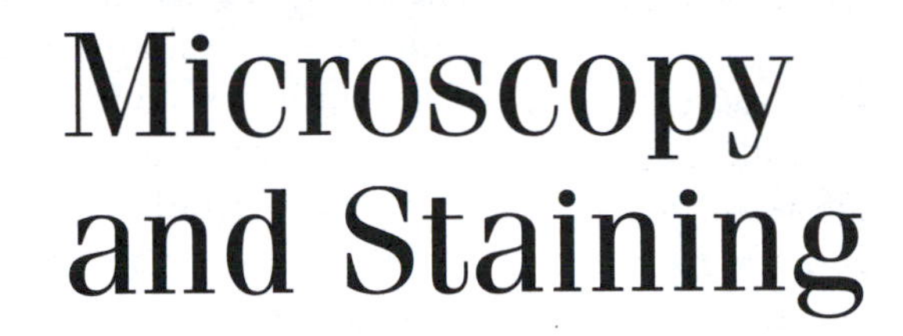

3 Microscopy and Staining

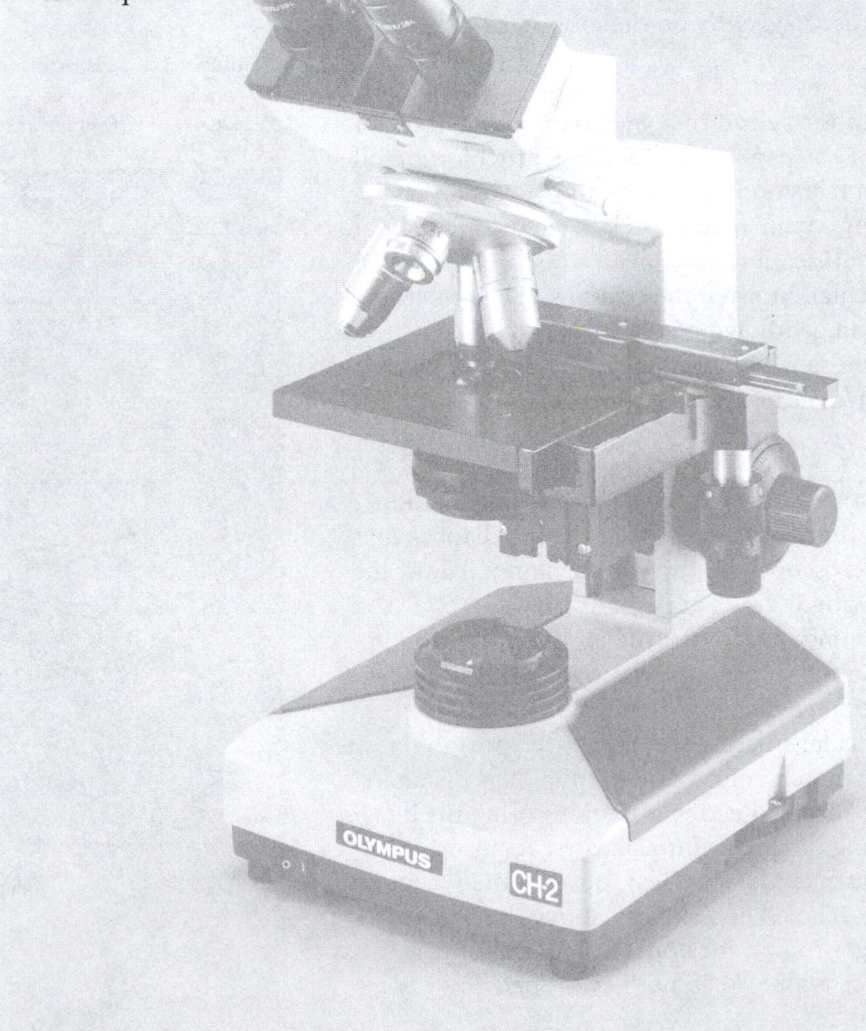

Microbiology as a biological discipline would not be what it is today without microscopes and cytological stains. Our ability to visualize, sometimes in great detail, the form and structure of microbes too small or transparent to otherwise be seen is due to developments in microscopy and staining techniques. In this section, you will learn (or refine) microscope skills. Then, you will learn simple and more sophisticated bacterial staining techniques.

3-1 INTRODUCTION TO THE LIGHT MICROSCOPE

INSTRUCTIONS FOR MICROSCOPE USE

Proper use of the microscope is essential for your success in microbiology. Fortunately, with practice and by following a few simple guidelines, you can achieve satisfactory results quickly. Since student labs may be supplied with a variety of microscopes, your instructor may supplement the following procedures and guidelines with instructions specific to your equipment. Refer to Figure 3-1 as you read the following (if working independently) or follow along on your microscope as your instructor guides you. (Note: This is a thorough treatment of microscope use and not all parts may be immediately relevant to your laboratory. Refer back to this exercise as necessary.)

Transport

- Carry your microscope to your workstation using both hands—one hand grasping the microscope's arm, the other supporting the microscope beneath its base.
- Place the microscope *gently* on the table.

Cleaning

- Use cotton swabs to clean the objective, ocular, and condenser lenses. Lens paper is useful for gently blotting oil from the oil immersion lens, but cotton swabs dipped in pure ethanol or commercial lens cleaning solution should be used for final cleaning.
- To clean an ocular, moisten the cotton swab with cleaning solution and gently wipe in a spiral motion starting at the center of the lens and working outward. Always wipe lenses dry with a clean cotton swab.

Operation

- Raise the substage condenser to its maximum position nearly even with the stage and open the iris diaphragm.
- Plug in the microscope and turn on the lamp. Adjust the light intensity slowly to its maximum.
- Move the scanning objective (usually 4X) into position.
- Place a slide on the stage in the mechanical slide holder. Center the specimen over the opening in the stage.
- If using a binocular microscope, adjust the position of the two oculars to match your own interpupillary distance.
- Use the coarse focus adjustment knob to bring the image into focus. Bring the image into sharpest focus using the fine focus adjustment knob. Then, observe the specimen with your eyes relaxed and slightly above the oculars to allow the images to fuse into one. If you are using a monocular microscope, keep both eyes open anyway to reduce eye fatigue.
- If you are using a binocular microscope, you may adjust the oculars' focus to compensate for differences in visual acuity of your two eyes.
- Adjust the iris diaphragm and condenser position to produce optimum illumination, contrast and image. (As a rule of thumb, use the maximum light intensity combined with the smallest aperture in the iris diaphragm that produces optimum illumination.)
- Scan the specimen to locate a promising region to examine in more detail.
- If you are observing a nonbacterial specimen, progress through the objectives until you see the degree of structural detail necessary for your purposes. You will need to adjust the fine focus and illumination for each objective. Before advancing to the next objective, be sure to position a desirable portion of the specimen in the center of the field or you risk "losing" it at the higher magnification.
- If you are working with a bacterial smear, you will need to use the oil immersion lens.
    - To use the oil immersion lens, work through the low (10X), then high dry (40X) objectives, adjusting the fine focus and illumination for each. Before advancing to the next objective, be sure to position a desirable

Figure 3-1 A BINOCULAR COMPOUND MICROSCOPE
A quality microscope is an essential tool for microbiologists. (Photograph courtesy of Olympus America Inc.)

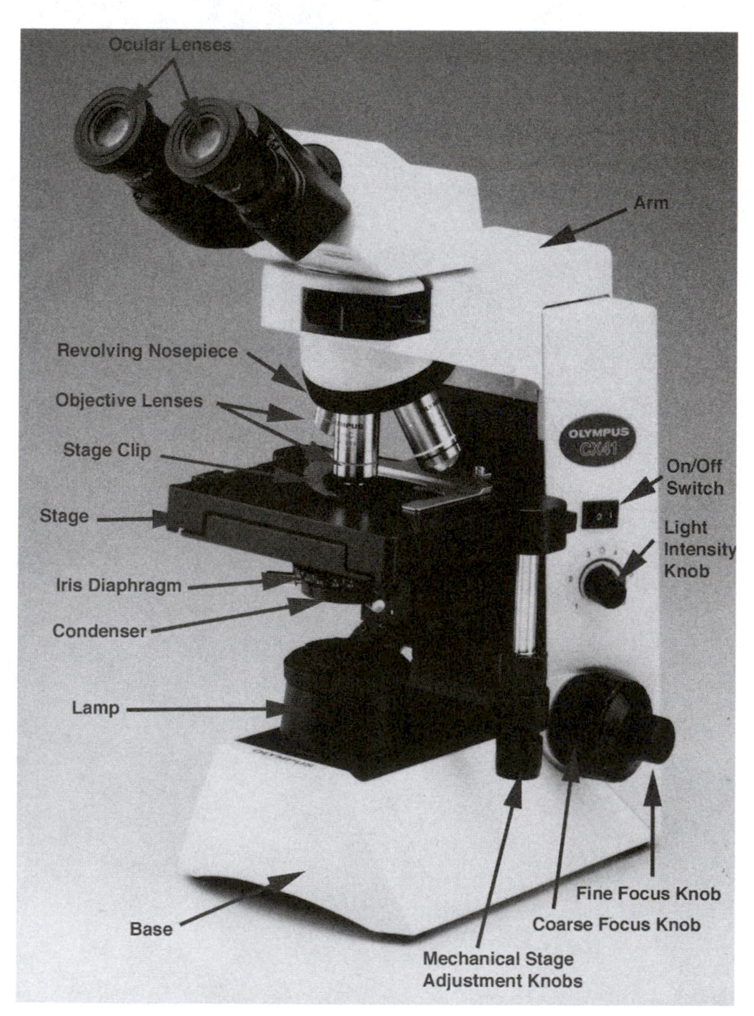

portion of the specimen in the center of the field or you risk "losing" it at the higher magnification.
- When the specimen is in focus under high dry, rotate the nosepiece to a position midway between the high dry and oil immersion lenses. Then, place a drop of immersion oil on the specimen. *Be careful not to get any oil on the microscope or its lenses, and be sure to clean it up if you do.*
- Rotate the oil lens so its tip is submerged in the oil drop.
- Focus and adjust the illumination to maximize image quality.
- **Note:** Do not move the stage down to add oil to the slide or the specimen will no longer be in focus. On a properly adjusted microscope the oil and the high dry lenses have the same focal plane. Therefore, when a specimen is in focus on high dry, the oil lens, although longer, will also be in focus and won't touch the slide when rotated into position.

- When finished, lower the stage (or raise the objective) and remove the slide. Dispose of the freshly prepared slides in a jar of disinfectant; return permanent slides to storage.

Storage

- When finished for the day, be sure to do the following:
  - Move the scanning objective into position.
  - Center the mechanical stage.
  - Lower the light intensity to its minimum, then turn off the light.
  - Wrap the electrical cord according to your particular lab rules.
  - Clean any oil off the lenses, stage, *etc.* Be sure to use only cotton swabs or lens paper for cleaning any of the optical surfaces of the microscope (see "Cleaning" on page 42).
  - Return the microscope to its appropriate storage place.

Photographic Atlas Reference
Microscopy Page 23

MATERIALS NEEDED FOR THIS EXERCISE

- Compound light microscope
- Nonsterile cotton swabs
- Lens cleaning solution or 95% ethanol

PROCEDURE

This Lab Exercise involves becoming familiar with your microscope. Actual examination of specimens will be done in subsequent Lab Exercises as assigned by your instructor.

1. Examine the ocular lenses on your microscope. Determine the magnification and record in the table below.
2. Examine the objective lenses on your microscope. Determine the magnification and numerical aperture of each, and record in the table below.
3. Use the formula below to calculate the total magnification for each objective lens. Record these in the table below.

$$\text{Total Magnification} = \text{Objective Lens Magnification} \times \text{Ocular Lens Magnification}$$

REFERENCES

Ash, Lawrence R. and Thomas C. Orihel. 1991. Pages 187–190 in *Parasites: A Guide to Laboratory Procedures and Identification.* American Society for Clinical Pathology (ASCP) Press, Chicago, IL.

Forbes, Betty A., Daniel F. Sahm, and Alice S. Weissfield. 2002. Pages 119–121 in *Bailey and Scott's Diagnostic Microbiology*, 11th Ed. Mosby, Inc., St. Louis, MO.

OBSERVATIONS AND INTERPRETATIONS

Record the relevant values off your microscope and perform the calculations of total magnification for each lens.

OBSERVATIONS AND INTERPRETATIONS

LENS SYSTEM	MAGNIFICATION OF OBJECTIVE LENS	MAGNIFICATION OF OCULAR LENS	TOTAL MAGNIFICATION	NUMERICAL APERTURE
Scanning				
Low Power				
High Dry				
Oil Immersion				
Condenser Lens				

3-2 CALIBRATION OF THE OCULAR MICROMETER

INTRODUCTION AND SAMPLE CALCULATIONS

An **ocular micrometer** is a type of ruler installed in the microscope eyepiece composed of uniform but unspecified graduations (Figure 3-2). As such, it must be calibrated before any viewed specimens can be measured. The device used to calibrate ocular micrometers is called a **stage micrometer.** As illustrated in Figure 3-3, a stage micrometer is a type of microscope slide containing a ruler with 10 µm and 100 µm graduations. (Other measuring instruments may be used in place of a stage micrometer, as in Figure 3-4.) When the stage micrometer is placed on the stage, it is magnified by the objective being used; therefore, the size of the graduations (relative to the ocular micrometer divisions) increases as magnification increases. Consequently, the *value* of ocular micrometer divisions decreases as magnification increases. For this reason, calibration must be performed for each magnification.

As shown in Figure 3-5, the stage micrometer is placed on the stage and brought into focus such that it

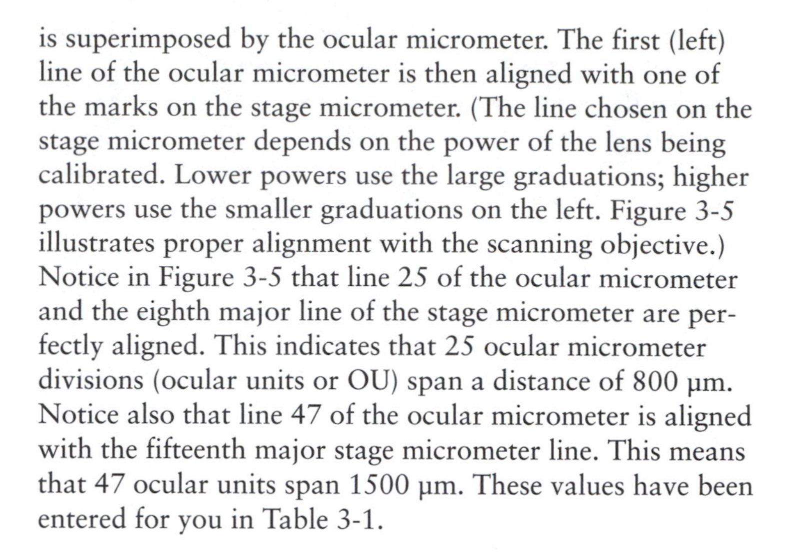
is superimposed by the ocular micrometer. The first (left) line of the ocular micrometer is then aligned with one of the marks on the stage micrometer. (The line chosen on the stage micrometer depends on the power of the lens being calibrated. Lower powers use the large graduations; higher powers use the smaller graduations on the left. Figure 3-5 illustrates proper alignment with the scanning objective.) Notice in Figure 3-5 that line 25 of the ocular micrometer and the eighth major line of the stage micrometer are perfectly aligned. This indicates that 25 ocular micrometer divisions (ocular units or OU) span a distance of 800 µm. Notice also that line 47 of the ocular micrometer is aligned with the fifteenth major stage micrometer line. This means that 47 ocular units span 1500 µm. These values have been entered for you in Table 3-1.

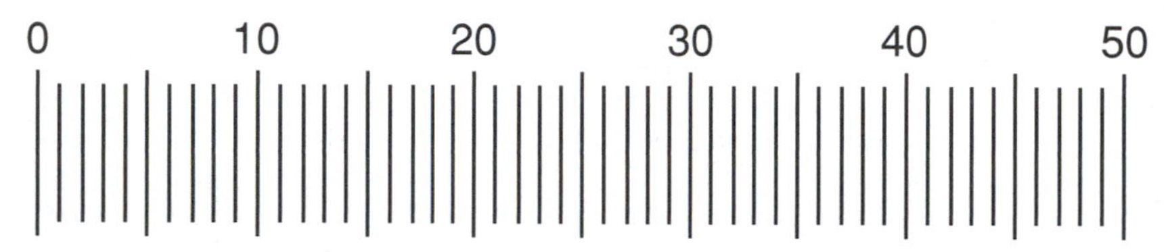

■ **Figure 3-2 An Ocular Micrometer**
The micrometer is a scale with uniform increments of unknown size.

■ **Figure 3-3 A Stage Micrometer**
This micrometer is 2200 µm long. The major divisions are 100 µm apart. The 200 µm at the left are divided into 10 µm increments.

■ **Figure 3-4 A Hemacytometer**
Any instrument with markings of known distance apart may be used as a stage micrometer. The hemacytometer is a grid made of lines 50 µm apart (lighter arrows). A larger grid is formed by lines 200 µm apart (darker arrows). Use any horizontal line as the micrometer and the smallest divisions are 50 µm apart.

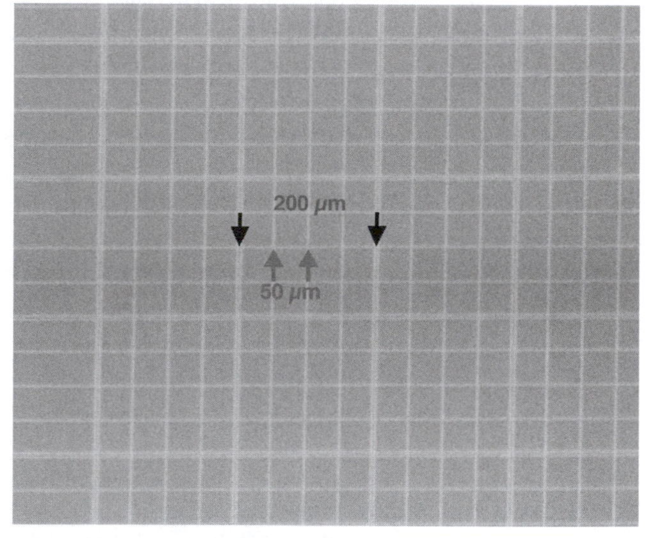

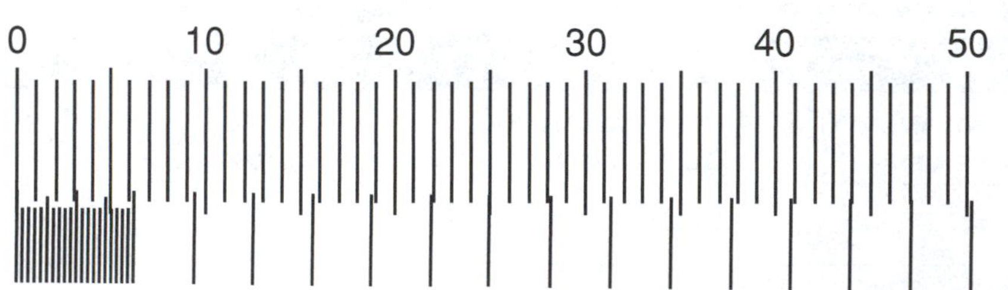

■ **Figure 3-5 This is What They Look Like in Use**
The stage micrometer as viewed through the microscope with the ocular micrometer superimposed on it.

TABLE 3-1 SAMPLE DATA FROM FIGURE 3-5

STAGE MICROMETER	OCULAR MICROMETER
800 µm	25 OU
1500 µm	47 OU

To determine the value of an ocular unit on a given magnification, divide the distance (from the stage micrometer) by the corresponding number of ocular units.

$$\frac{800\ \mu m}{25\ \text{ocular units}} = 32\frac{\mu m}{OU}$$

$$\frac{1500\ \mu m}{47\ \text{ocular units}} = 32\frac{\mu m}{OU}$$

As shown in this example, it is customary to record more than one measurement. Each measurement is calculated separately. If the calculated ocular unit values differ, use their arithmetic mean as the calibration for that objective lens.

As mentioned previously, each magnification must be calibrated. Calibrating the oil immersion lens may be difficult to accomplish using the stage micrometer. If this is the case, its value can be calculated using the calibration value of one of the other lenses. Refer to Table 3-2 for the magnifications of each lens system of a typical microscope. Notice in the table that the magnification of the oil immersion lens is ten times greater than the low power lens. This means that objects viewed on the stage (stage micrometer *or* specimens) become ten times larger when changing from low power to oil immersion. But, because the magnification of the ocular micrometer does not change, an ocular division now covers only one-tenth the distance. Thus, the size of an ocular unit using the oil immersion lens can be calculated by dividing the calibration for low power by 10.

Ocular micrometer values can be calculated for any lens using values from any other lens and provide a good check of measured values. For practice, assuming the calibration for the scanning objective is 32 µm/OU, calculate the low, high dry, and oil immersion calibrations. Write the values in Table 3-3. (Remember that the calculated size of an ocular unit for the scanning objective was 32 µm.)

Once you have determined the ocular unit values for each objective lens, use the ocular micrometer as a ruler to measure specimens. For instance, if you determine that under the scanning objective a cell is 5 ocular units long, the cell's actual length would be determined as follows (using the sample values from Table 3-3):

Cell Dimension = Ocular Units × Calibration

Cell Dimension = 5 Ocular Units × 32 µm/OU

Cell Dimension = 160 µm

TABLE 3-2 MAGNIFICATIONS OF A TYPICAL MICROSCOPE

Power	Objective Lens	Ocular Lens	Total Magnification
Scanning	4X	10X	4X x 10X = 40X
Low Power	10X	10X	10X x 10X = 100X
High Dry Power	40X	10X	40X x 10X = 400X
Oil Immersion	100X	10X	100X x 10X =1000X

TABLE 3-3 SAMPLE PROBLEM: CALCULATED CALIBRATION VALUES FROM DATA IN FIGURE 3-5

Total Magnification	Calibration (µm/OU)
40X	32 µm/OU
100X	
400X	
1000X	

Materials Needed For This Exercise

Per Student

- Compound microscope equipped with an ocular micrometer
- Stage micrometer

Procedure

This Lab Exercise involves calibrating the ocular micrometer on your microscope. Actual measurement of specimens will be done in subsequent Lab Exercises as assigned by your instructor.

1. Check your microscope and determine which ocular has the micrometer in it.
2. Move the scanning objective into position.
3. Place the stage micrometer on the stage and position it so that its image is superimposed by the ocular micrometer and their left-hand marks line up.
4. Examine the two micrometers and, as described above, record two or three points where they line up exactly. Record these values in the table below and calculate the value of each ocular unit.

SCANNING OBJECTIVE LENS

Stage Micrometer (µm)	Ocular Micrometer (OU)	Calibration (µm/OU)

5. Change to medium power and repeat the process.

LOW POWER OBJECTIVE LENS		
STAGE MICROMETER (µM)	**OCULAR MICROMETER (OU)**	**CALIBRATION (µM/OU)**

6. Change to high dry power and repeat the process.

HIGH DRY OBJECTIVE LENS		
STAGE MICROMETER (µM)	**OCULAR MICROMETER (OU)**	**CALIBRATION (µM/OU)**

7. Change to the oil immersion lens and repeat the process. If this cannot be accurately done, complete the calibration from the value of another lens.

OIL IMMERSION OBJECTIVE LENS		
STAGE MICROMETER (µM)	**OCULAR MICROMETER (OU)**	**CALIBRATION (µM/OU)**

8. Compute average calibrations for each objective lens and record these in the table below.

AVERAGE CALIBRATIONS FOR MY MICROSCOPE	
OBJECTIVE LENS	**AVERAGE CALIBRATION (µM/OU)**
Scanning	
Low Power	
High Dry Power	
Oil Immersion	

9. As long as you keep this microscope all term, you may use your calibrations in the table without recalibrating it.

REFERENCES

Abramoff, Peter and Robert G. Thompson. 1982. Pages 5 and 6 in *Laboratory Outlines in Biology—III*. W. H. Freeman and Company. San Francisco, CA.

Ash, Lawrence R. and Thomas C. Orihel. 1991. Pages 187–190 in *Parasites: A Guide to Laboratory Procedures and Identification*. American Society for Clinical Pathology (ASCP) Press, Chicago, IL.

3-3 EXAMINATION OF EUKARYOTIC MICROBES

Most of this manual is devoted to prokaryotes, but in this exercise you will be given the opportunity to examine various eukaryotic microorganisms—protozoans, algae, and fungi. (Please refer to your text or the *Photographic Atlas* for a description of common eukaryotic groups.) This will not only serve to familiarize you with simple eukaryotes, but also give you practice at using the microscope, measuring specimens, and making wet-mount preparations.

Photographic Atlas References
Protozoans Page 175 and Fungi Page 159

MATERIALS NEEDED FOR THIS EXERCISE

Per Class

- Prepared slides or living cultures of a variety of protists and fungi (such as *Amoeba, Paramecium, Leishmania*—prepared only, *Spirogyra, Volvox,* diatoms, and *Saccharomyces.*)

Per Student

- Clean glass slides and cover glasses
- Compound microscope
- Cytological stains (*e.g.,* methylene blue, I_2KI)
- Methyl cellulose
- Immersion oil
- Cotton swabs
- Lens paper
- Lens cleaning solution

PROCEDURE

1. Obtain a microscope and place it on the table or workspace. Check to be sure the stage is all the way down and the scanning objective is in place.
2. Begin with a prepared slide. Clean it with a Kim-wipe tissue if it is dirty, then place it on the microscope stage. Center the specimen under the scanning objective.
3. Follow the instructions given in Exercise 3-1, to bring the specimen into focus at the highest magnification that allows you to see the entire structure you want to view.
4. Practice scanning with the mechanical stage until you are satisfied that you have seen everything interesting to see. Sketch what you see in the table provided on page 48.
5. Measure cellular dimensions and record these in the table provided.
6. Repeat with as many slides as you have time for.
7. Prepare wet mounts of available specimens by following the Procedural Diagram in Figure 3-6. Sketch what you see and record cellular dimensions in the table provided.

■ Figure 3-6 Wet Mount Procedural Diagram
Wet mounts are made using living specimens.

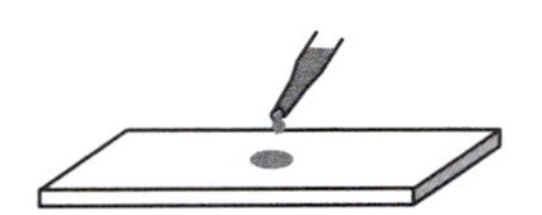

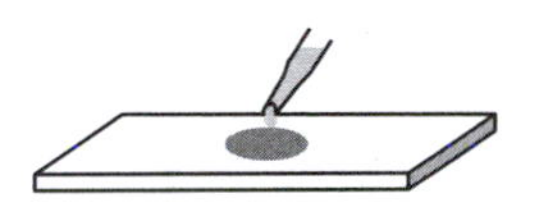

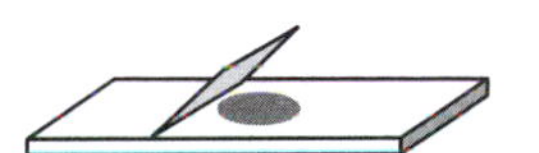

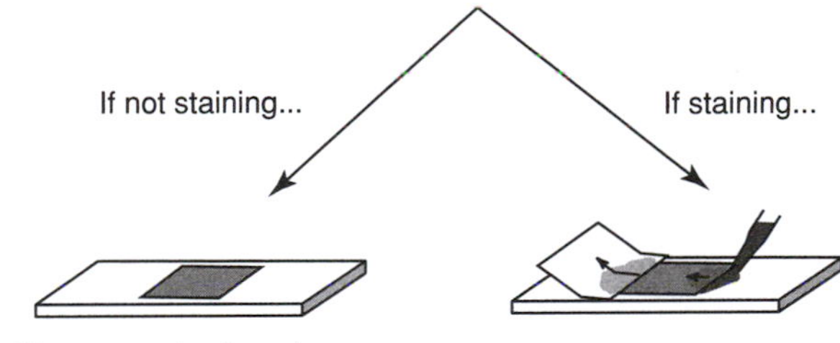

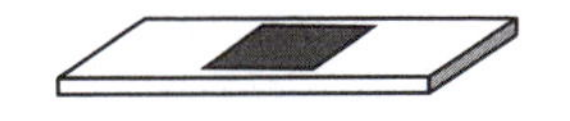

8. When you are finished observing specimens, blot the oil from the oil immersion lens (if used) with a lens paper and do a final cleaning with a cotton swab and alcohol or lens cleaning solution. Dry the lenses with a clean cotton swab.
9. Return all lenses and adjustments to their storage positions before putting the microscope away.

REFERENCES

Campbell, Neil A. and Jane B. Reece, and Lawrence G. Mitchell. 2002. Chapter 28 in *Biology*, 6th Ed. Benjamin/Cummings Publishing Company, Inc., San Francisco, CA.

Guttman, Burton S. 1999. Chapter 29 in *Biology*. WCB/McGraw-Hill, Boston.

OBSERVATIONS AND INTERPRETATIONS

As you observe specimens, fill in the table below.

OBSERVATIONS AND INTERPRETATIONS

ORGANISM (INCLUDE WET MOUNT OR PREPARED SLIDE)	SKETCH (INCLUDE MAGNIFICATION AND STAIN)	DIMENSIONS	IDENTIFYING CHARACTERISTICS

Bacterial Structure and Simple Stains

Cytoplasm is transparent, so viewing cells with the light microscope is difficult without the aid of stains to provide contrast. In this set of exercises, you will learn how to correctly prepare a bacterial smear for staining and how to perform simple and negative stains. Cell morphology, size, and arrangement may then be determined. In a medical laboratory, these are usually determined with a Gram stain (Exercise 3-6), but you will be using simple stains as an introduction to the staining process as well as an introduction to these cellular characteristics.

3-4 SIMPLE STAINS

Photographic Atlas Reference
Simple Stains Page 27

MATERIALS NEEDED FOR THIS EXERCISE

Per Student

- Clean glass microscope slides
- Methylene blue stain
- Safranin stain
- Crystal violet stain
- Disposable gloves
- Staining tray
- Staining screen
- Bibulous paper tablet
- Slide holder
- Recommended organisms:
  - *Micrococcus luteus*
  - *Staphylococcus epidermidis*
  - *Bacillus cereus*
  - *Rhodospirillum rubrum*
  - *Vibrio harveyi*
  - *Moraxella catarrhalis*

PROCEDURE

1. A bacterial smear (emulsion) is made prior to most staining procedures. Follow the Procedural Diagram in Figure 3-7 to prepare bacterial smears of each organism.
2. Heat-fix each smear as described in Figure 3-7.
3. Follow the basic staining procedure illustrated in the Procedural Diagram in Figure 3-8. Prepare two slides with each stain using the following times:
   Crystal violet: stain for 30 to 60 seconds
   Safranin: stain for up to 1 minute
   Methylene blue: stain for 30 to 60 seconds
   Record your actual staining times in the table provided so you can adjust for over staining or under staining.
4. Observe each slide using the oil immersion lens. Record your observations of cell morphology, arrangement, and size in the table provided.
5. Dispose of the slides and used stain according to your laboratory's policy.

■ **Figure 3-7 Procedural Diagram—Making a Bacterial Smear (Emulsion)**
Preparation of uniform bacterial smears will make consistent staining results easier to obtain. Heat-fixing the smear kills the bacteria, makes them adhere to the slide, and coagulates their protein for better staining. **Caution: Avoid producing aerosols. Do not spatter the smear as you mix it, do not blow on or wave the slide to speed-up air-drying, and do not overheat when heat-fixing.**

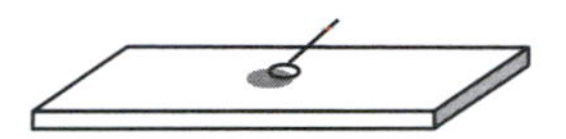

1. Place a small drop of water (not too much) on a clean slide using an inoculating loop.

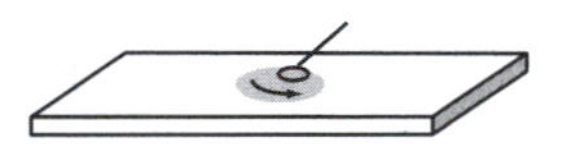

2. Aseptically add bacteria to the water. Mix in the bacteria and spread the drop out. Avoid spattering the emulsion as you mix. Flame your loop when done.

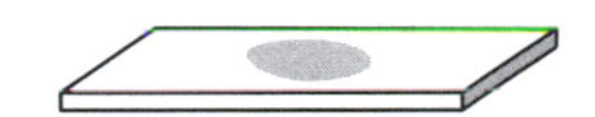

3. Allow the smear to air dry. If prepared correctly, the smear should be slightly cloudy.

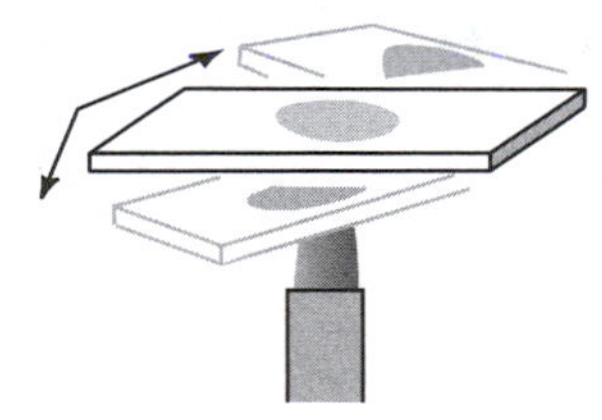

4. Using a slide holder, pass the smear through the upper part of a flame two or three times. This heat-fixes the preparation. Avoid overheating the slide as aerosols may be produced.

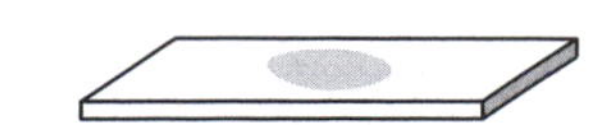

5. Allow the slide to cool, then continue with the staining protocol.

■ Figure 3-8 Procedural Diagram—Simple Stain.
Staining times differ for each stain, but cell density of your smear also affects staining time. Strive for consistency in making your smears.
Caution: Be sure to flame your loop after cell transfer and properly dispose of the slide when you are finished observing it.

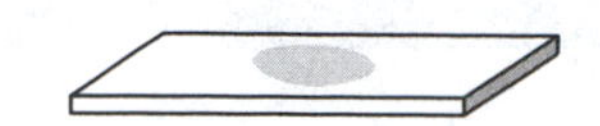

1. Begin with a heat-fixed emulsion.

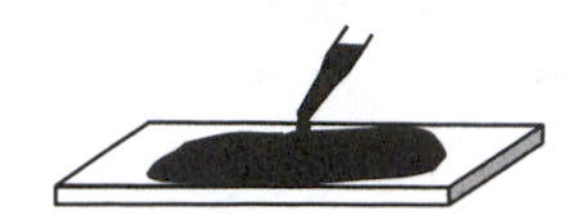

2. Cover the smear with stain.
Use a staining tray to catch excess stain.

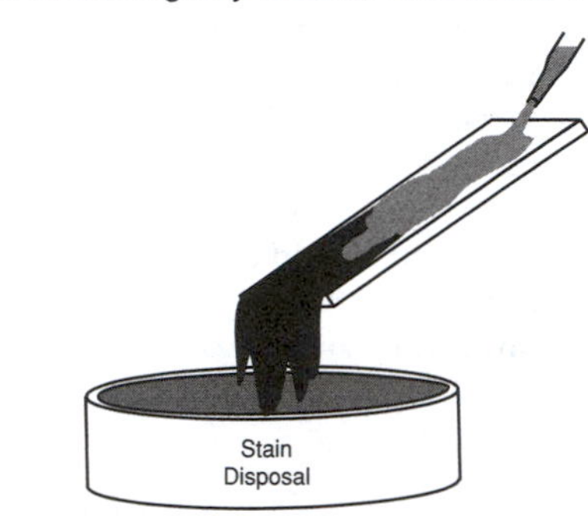

3. Grasp the slide with a slide holder.
Rinse the slide with distilled water.
Dispose of the excess stain according to your lab practices.

4. Gently blot dry in a tablet of bibulous paper.
Do not rub.
Observe under oil immersion.

REFERENCES

Chapin, Kimberle. 1995. Chapter 4 in *Manual of Clinical Microbiology*, 6th Ed., edited by Patrick R. Murray, Ellen Jo Baron, Michael A. Pfaller, Fred C. Tenover, and Robert H. Yolken. American Society for Microbiology, Washington, DC.

Chapin, Kimberle C. and Patrick R. Murray. 2003. Pages 257–259 in *Manual of Clinical Microbiology*, 8th Ed., edited by Patrick R. Murray, Ellen Jo Baron, James H. Jorgensen, Michael A. Pfaller, and Robert H. Yolken. American Society for Microbiology, Washington, DC.

Claus, G. William. 1989. Chapter 5 in *Understanding Microbes—A Laboratory Textbook for Microbiology*. W. H. Freeman and Company, New York, NY.

DIFCO Laboratories. 1984. *DIFCO Manual*, 10th Ed. DIFCO Laboratories, Detroit, MI.

Forbes, Betty A., Daniel F. Sahm, and Alice. S. Weissfeld. 2002. Chapter 9 in *Bailey and Scott's Diagnostic Microbiology*, 11th Ed. Mosby-Yearbook, Inc., St. Louis, MO.

Murray, R. G. E., Raymond N. Doetsch, and C. F. Robinow. 1994. Page 27 in *Methods for General and Molecular Bacteriology*, edited by Philipp Gerhardt, R. G. E. Murray, Willis A. Wood, and Noel R. Krieg. American Society for Microbiology, Washington, DC.

Norris, J. R. and Helen Swain. 1971. Chapter II in *Methods in Microbiology*, Volume 5A, edited by J. R. Norris and D. W. Ribbons. Academic Press, Ltd., London.

Power, David A. and Peggy J. McCuen. 1988. Page 4 in *Manual of BBL® Products and Laboratory Procedures*, 6th Ed. Becton Dickinson Microbiology Systems, Cockeysville, MD.

■ OBSERVATIONS AND INTERPRETATIONS

Record your observations in the table below.

OBSERVATIONS AND INTERPRETATIONS			
ORGANISM	STAIN AND DURATION	CELLULAR MORPHOLOGY AND ARRANGEMENT (INCLUDE A SKETCH)	CELL DIMENSIONS

3-5 NEGATIVE STAIN

Photographic Atlas Reference
Negative Staining Page 28

MATERIALS NEEDED FOR THIS EXERCISE

Per Student

- Nigrosin stain or eosin stain
- Clean glass microscope slides
- Disposable gloves
- Recommended organisms:
  - *Micrococcus luteus*
  - *Bacillus megaterium*
  - *Rhodospirillum rubrum*

PROCEDURE

1. Follow the Procedural Diagram in Figure 3-9 to prepare a negative stain of each organism.
2. Dispose of the spreader slide in a disinfectant jar immediately after use.
3. Observe using the oil immersion lens. Record your observations in the table provided.
4. Dispose of the specimen slide in a disinfectant jar after use.

REFERENCES

Claus, G. William. 1989. Chapter 5 in *Understanding Microbes—A Laboratory Textbook for Microbiology*. W. H. Freeman and Company, New York, NY.

Murray, R. G. E., Raymond N. Doetsch and C. F. Robinow. 1994. Page 27 in *Methods for General and Molecular Bacteriology*, edited by Philipp Gerhardt, R. G. E. Murray, Willis A. Wood, and Noel R. Krieg. American Society for Microbiology, Washington, DC.

Figure 3-9 Procedural Diagram—Negative Stain
Be sure to sterilize your loop after transfer and to appropriately dispose of the spreader and specimen slides.

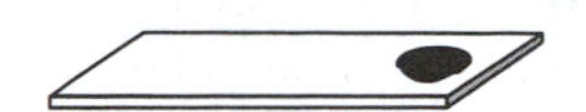

1. Begin with a drop of acidic stain at one end of a clean slide.

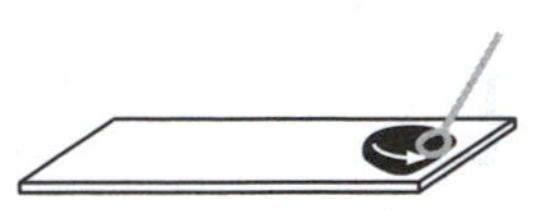

2. Aseptically add organisms and emulsify with a loop. Do not over-inoculate and avoid spattering the mixture Sterilize the loop after emulsifying.

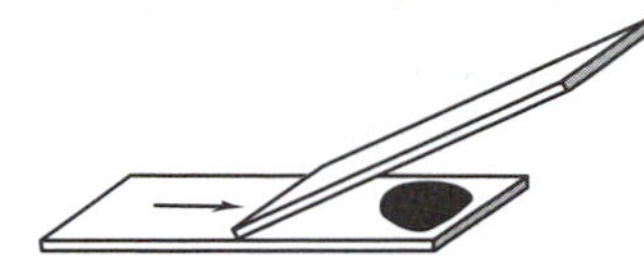

3. Take a second clean slide, place it on the surface of the first slide, and draw it back into the drop.

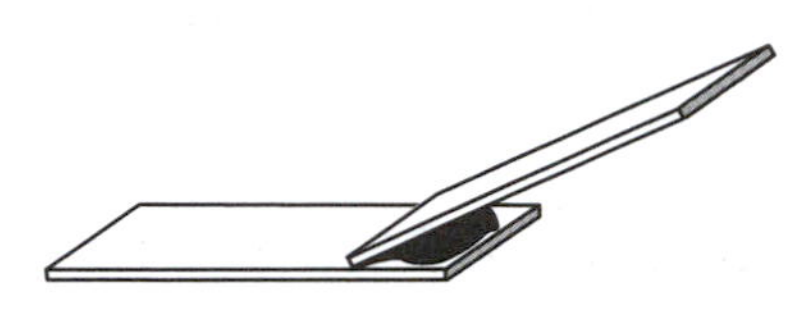

4. When the drop flows across the width of the spreader slide...

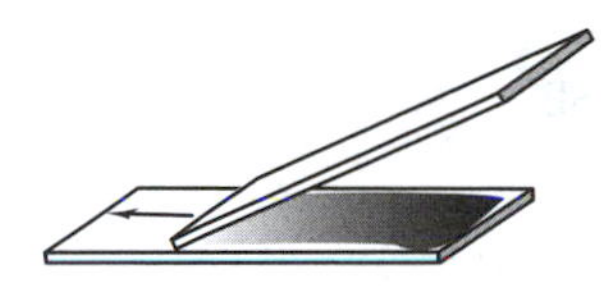

5. ...push the spreader slide to the other end. Dispose of the spreader slide in a jar of disinfectant.

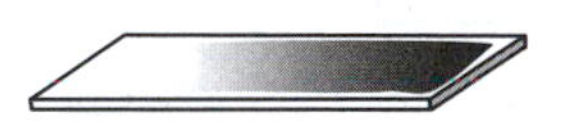

6. Air dry and observe under the microscope. Do NOT heat fix.

OBSERVATIONS AND INTERPRETATIONS

Record your observations in the table below.

OBSERVATIONS AND INTERPRETATIONS		
ORGANISM	CELLULAR MORPHOLOGY AND ARRANGEMENT (INCLUDE A SKETCH)	CELL DIMENSIONS

Differential Stains

Differential stains allow a microbiologist to detect differences between organisms or differences between parts of the same organism. In practice, these are used much more frequently than simple stains because they not only allow determination of cell size, shape, and arrangement (as with a simple stain), but provide information about other features as well. The Gram stain is the most commonly used differential stain in bacteriology. Other differential stains are used for organisms not distinguishable by the Gram stain or for those possessing other important cellular attributes, such as acid-fastness, a capsule, spores, or flagella. With the exception of the acid-fast stain, these other stains are sometimes referred to as "structural stains."

3-6 GRAM STAIN

Photographic Atlas Reference
Gram Stain Page 35

MATERIALS NEEDED FOR THIS EXERCISE

Per Student

- Clean glass microscope slides
- Sterile toothpick
- Gram Stain Solutions (Commercial kits are available)
  - Gram crystal violet
  - Gram iodine
  - 95% ethanol (or ethanol/acetone solution)
  - Gram safranin
- Bibulous paper
- Disposable gloves
- Staining tray
- Staining screen
- Slide holder
- Recommended organisms (overnight cultures grown on agar slants):
  - *Staphylococcus epidermidis*
  - *Escherichia coli*
  - *Moraxella catarrhalis*
  - *Bacillus subtilis*

PROCEDURE

1. Follow the procedure illustrated in Figure 3-7 to prepare and heat-fix smears of *Staphylococcus epidermidis* and *Escherichia coli* right next to one another on the same clean glass slide. (If you make the emulsions at opposite ends of the slide, you may find it difficult to stain and decolorize each equally.) Strive for preparing smears of uniform thickness. Thick smears risk being under decolorized.
2. Repeat step one for *Moraxella catarrhalis* and *Bacillus subtilis*.
3. Since Gram stains require much practice, you may want to prepare several slides of each combination and let them be air-drying simultaneously. Then they'll be ready if you need them.
4. Use the sterile toothpick to obtain a sample from your teeth at the gum line. (Do not draw blood! What you want is easily removed from your gingival pockets.) Transfer the sample to a drop of water on a clean glass slide, air dry, and heat fix.
5. Follow the basic staining procedure illustrated in Figure 3-10. We recommend staining the pure cultures first. When your technique is consistent, then stain the oral sample.
6. Observe using the oil immersion lens. Record your observations of cell morphology and arrangement, dimensions, and Gram reactions in the table provided.
7. Dispose of the specimen slides in a jar of disinfectant after use.

REFERENCES

Chapin, Kimberle C. and Patrick R. Murray. 2003. Pages 258–260 in *Manual of Clinical Microbiology,* 8th Ed., edited by Patrick R. Murray, Ellen Jo Baron, James H. Jorgensen, Michael A. Pfaller, and Robert H. Yolken. American Society for Microbiology, Washington, DC.

Forbes, Betty A., Daniel F. Sahm, and Alice. S. Weissfeld. 2002. Chapter 9 in *Bailey and Scott's Diagnostic Microbiology*, 11th Ed. Mosby-Yearbook, Inc. St., Louis, MO.

Koneman, Elmer W., Stephen D. Allen, William M. Janda, Paul C. Schreckenberger, and Washington C. Winn, Jr. 1997. Chapter 14 in *Color Atlas and Textbook of Diagnostic Microbiology*, 5th Ed. J. B. Lippincott Company, Philadelphia, PA.

Murray, R. G. E., Raymond N. Doetsch and C. F. Robinow. 1994. Pages 31 and 32 in *Methods for General and Molecular Bacteriology*, edited by Philipp Gerhardt, R. G. E. Murray, Willis A. Wood, and Noel R. Krieg. American Society for Microbiology, Washington, DC.

Norris, J. R. and Helen Swain. 1971. Chapter II in *Methods in Microbiology,* Volume 5A, edited by J. R. Norris and D. W. Ribbons. Academic Press, Ltd., London.

Power, David A. and Peggy J. McCuen. 1988. Page 261 in *Manual of BBL® Products and Laboratory Procedures*, 6th Ed. Becton Dickinson Microbiology Systems, Cockeysville, MD.

■ **Figure 3-10 Procedural Diagram–Gram Stain**

Pay careful attention to the staining times. If your preparations do not give "correct" results, the most likely source of error is in the decolorization step. Adjust its timing accordingly on subsequent stains.

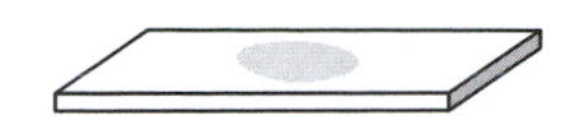

1. Begin with a heat-fixed emulsion.

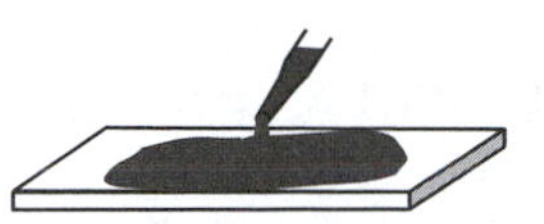

2. Cover the smear with Crystal Violet stain for 1 minute. Use a staining tray to catch excess stain.

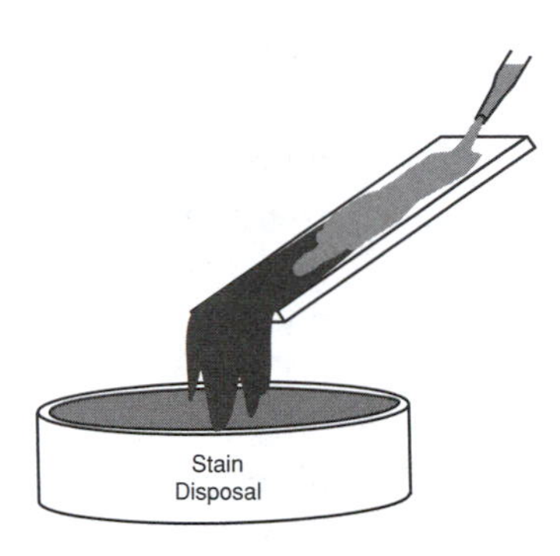

3. Grasp the slide with a slide holder. Gently rinse the slide with distilled water.

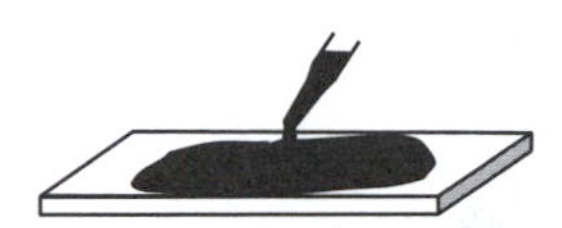

4. Cover the smear with Iodine stain for 1 minute. Use a staining tray to catch excess stain.

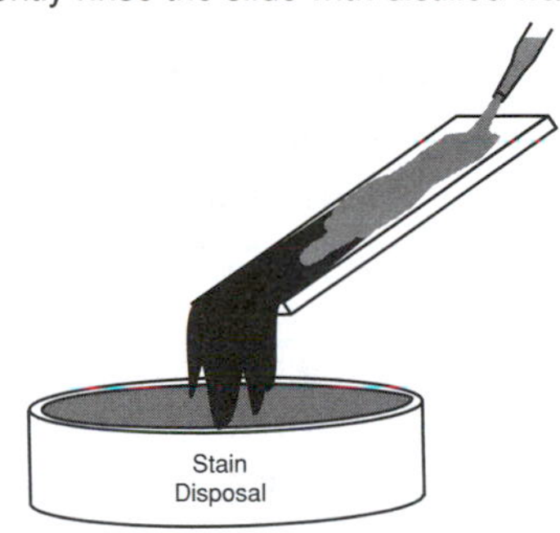

5. Grasp the slide with a slide holder. Gently rinse the slide with distilled water.

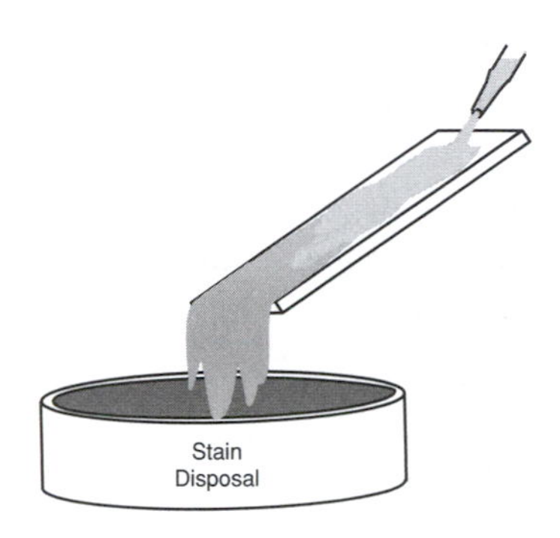

6. Decolorize with 95% ethanol or ethanol/acetone until the run-off is clear. Gently rinse the slide with distilled water.

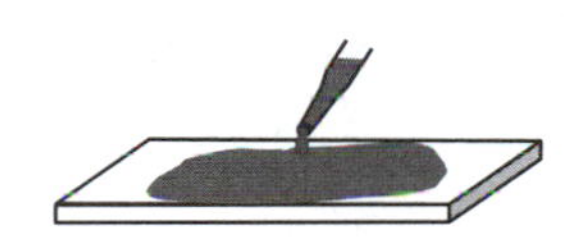

7. Counterstain with Safranin stain for 1 minute. Rinse with water.

8. Gently blot dry in a tablet of bibulous paper. Do not rub. Observe under oil immersion.

■ Observations and Interpretations

Record your observations in the table below.

Observations and Interpretations			
Organism or Source	Cellular Morphology and Arrangement (Include a Sketch)	Cell Dimensions	Gram Reaction (+/-)

3-7 ACID-FAST STAINS

Photographic Atlas Reference
Acid Fast Stains Page 38

MATERIALS NEEDED FOR THIS EXERCISE

Per Student

- Clean glass microscope slides
- Staining tray
- Staining screen
- Bibulous paper
- Slide holder
- Ziehl-Neelsen Stains (Complete kits are commercially available)
  - Methylene blue stain
  - Ziehl's carbolfuchsin stain
  - Acid alcohol (95% ethanol + 3% HCl)
- Kinyoun Stains (Complete kits are commercially available)
  - Kinyoun carbolfuchsin
  - Acid alcohol (95% ethanol + 3% HCl)
  - Brilliant green stain
- Sheep serum
- Heating apparatus (steam or hot plate)
- Nonsterile Petri dish for transporting slides
- Disposable gloves
- Lab coat or apron
- Eye goggles
- Recommended organisms:
  - *Mycobacterium phlei*
  - *Staphylococcus epidermidis*

PROCEDURE—ZIEHL-NEELSEN (ZN) METHOD

1. Prepare a smear of each organism on a clean glass slide as illustrated in Figure 3-7, substituting a drop of sheep serum for the drop of water. Air-dry and then heat-fix the smears. NOTE: you may make two separate smears right next to one another on the slide or mix the two organisms in one smear.
2. Follow the staining protocol shown in the Procedural Diagram (Figure 3-11). Use a steaming apparatus (such as the one in Figure 3-12) to heat the slide. If the slide must be carried to and from the steaming apparatus, put it in a covered Petri dish.
3. Observe using the oil immersion lens. Record your observations of cell morphology and arrangement, dimensions, and acid-fast reaction in the table below.
4. When finished, dispose of slides in a disinfectant jar.

PROCEDURE—KINYOUN METHOD

1. Prepare a smear of each organism on a clean glass slide as illustrated in Figure 3-7, substituting a drop of sheep serum for the drop of water. Air-dry and then heat-fix the smears. NOTE: you may make two separate smears right next to one another on the slide or mix the two organisms in one smear.
2. Follow the staining protocol shown in the Procedural Diagram (Figure 3-13).
3. Observe using the oil immersion lens. Record your observations of cell morphology and arrangement, dimensions, and acid-fast reaction in the table below.
4. When finished, dispose of slides in a disinfectant jar.

■ OBSERVATIONS AND INTERPRETATIONS

Record your observations in the table below.

OBSERVATIONS AND INTERPRETATIONS

ORGANISM	STAINING METHOD	CELLULAR MORPHOLOGY AND ARRANGEMENT (INCLUDE A SKETCH)	CELL DIMENSIONS	ACID-FAST REACTION (+/-)

■ **Figure 3-11 Procedural Diagram—ZN Acid-Fast Stain**
Be sure to perform this stain in a fume hood or well-ventilated area.

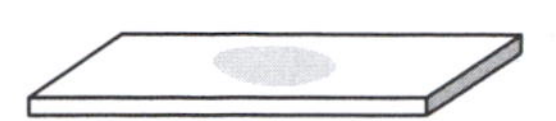

1. Begin with a heat-fixed emulsion.

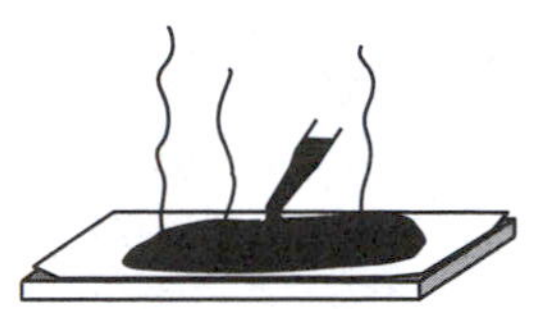

2. Cover the smear with a strip of bibulous paper.
Apply ZN carbolfuchsin stain.
Steam (as shown in Figure 3-12) for 5 minutes.
Keep the paper moist with stain.
Perform this step with adequate ventilation
and eye protection.
Do not boil the stain.

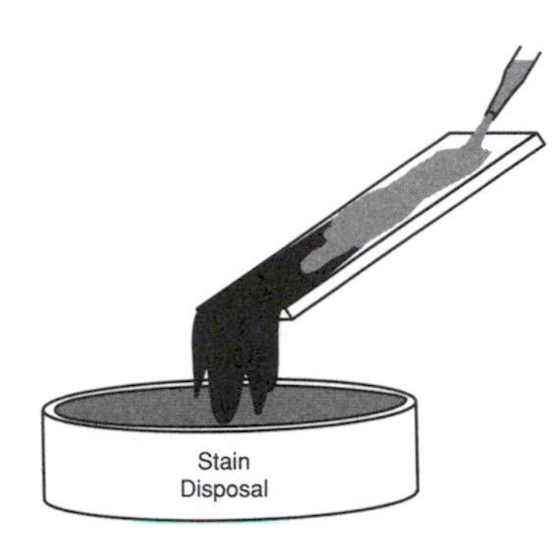

3. Grasp the slide with a slide holder.
Remove the paper and dispose of it properly.
Gently rinse the slide with distilled water.

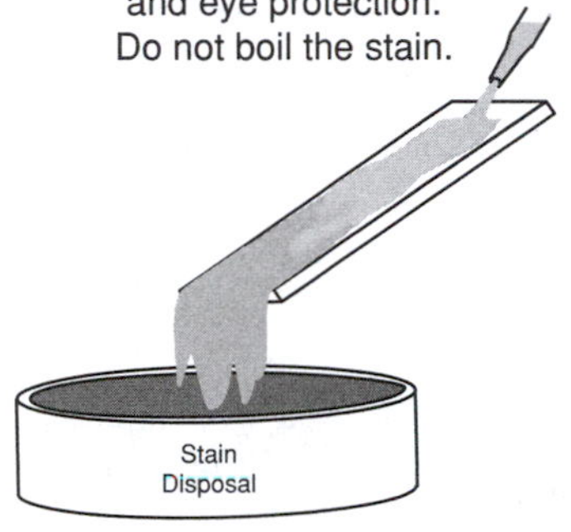

4. Continue holding the slide with a slide holder.
Decolorize with acid-alcohol (CAUTION!)
until the run-off is clear.
Gently rinse the slide with distilled water.

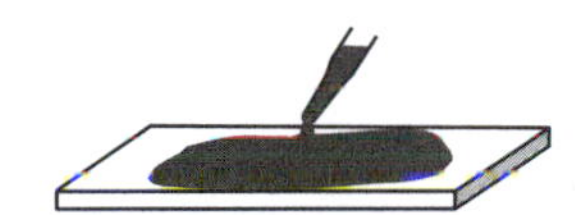

5. Counterstain with Methylene Blue stain for 1 minute.
Rinse with water.

6. Gently blot dry in a tablet of bibulous paper.
Do not rub.
Observe under oil immersion.

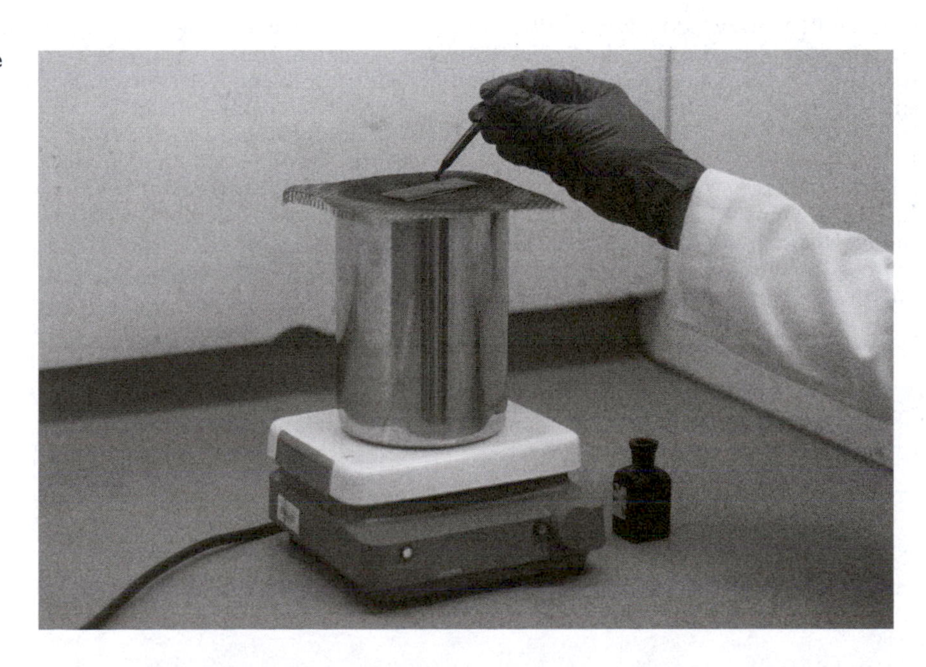

■ **Figure 3-12 Steaming the Slide During the Ziehl-Neelsen Procedure**
Carefully steam the slide to force the carbolfuchsin into acid-fast cells. Do not boil the slide or let it dry out. Keep it moist with stain for the entire five minutes of steaming. Caution: this should be performed in a well-ventilated area with hand, clothing and eye protection.

■ Figure 3-13 Procedural Diagram—Kinyoun Acid-Fast Stain

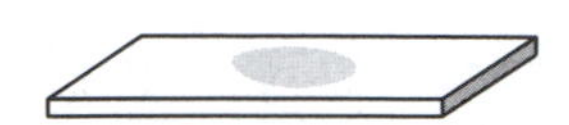

1. Begin with a heat-fixed emulsion.

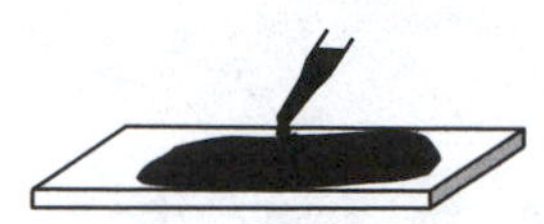

2. Apply Kinyoun carbolfuchsin stain for 5 minutes. Perform this step with adequate ventilation.

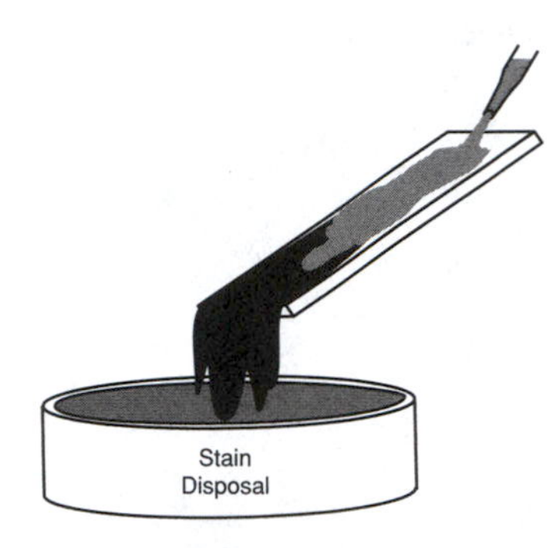

3. Grasp the slide with a slide holder. Gently rinse the slide with distilled water.

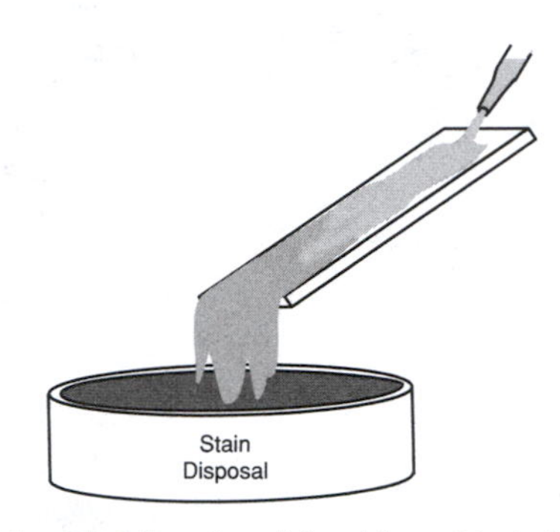

4. Continue holding the slide with a slide holder. Decolorize with acid-alcohol (CAUTION!) until the run-off is clear. Gently rinse the slide with distilled water.

5. Counterstain with Brilliant Green stain for 1 minute. Rinse with water.

6. Gently blot dry in a tablet of bibulous paper. Do not rub. Observe under oil immersion.

REFERENCES

Chapin, Kimberle C. and Patrick R. Murray. 2003. Pages 259–261 in *Manual of Clinical Microbiology,* 8th Ed., edited by Patrick R. Murray, Ellen Jo Baron, James H. Jorgensen, Michael A. Pfaller, and Robert H. Yolken. American Society for Microbiology, Washington, DC.

Forbes, Betty A., Daniel F. Sahm, and Alice. S. Weissfeld. 2002. Chapter 9 in *Bailey and Scott's Diagnostic Microbiology*, 11th Ed. Mosby-Yearbook, Inc., St. Louis, MO.

Murray, R. G. E., Raymond N. Doetsch, and C. F. Robinow. 1994. Page 32 in *Methods for General and Molecular Bacteriology*, edited by Philipp Gerhardt, R. G. E. Murray, Willis A. Wood, and Noel R. Krieg. American Society for Microbiology, Washington, DC.

Norris, J. R. and Helen Swain. 1971. Chapter II in *Methods in Microbiology*, Volume 5A, edited by J. R. Norris and D. W. Ribbons. Academic Press, Ltd., London.

Power, David A. and Peggy J. McCuen. 1988. Page 5 in *Manual of BBL® Products and Laboratory Procedures*, 6th Ed. Becton Dickinson Microbiology Systems, Cockeysville, MD.

3-8 CAPSULE STAIN

Photographic Atlas Reference
Capsule Stain Page 40

MATERIALS NEEDED FOR THIS EXERCISE

Per Student

- Clean glass slides
- Sheep serum
- Maneval's stain
- Congo red stain
- Staining tray
- Staining screen
- Bibulous paper tablet
- Slide holder
- Disposable gloves
- Sterile toothpicks
- Recommended organisms (18 to 24 hour skim milk pure cultures):
  - *Flavobacterium capsulatum (Sphingomonas capsulata)*
  - *Aeromonas hydrophila*
  - *Lactococcus lactis*

PROCEDURE

1. Follow the protocol in the Procedural Diagram (Figure 3-14) to make stains of the organisms supplied. Each specimen should be done on a separate slide. These specimens are not heat-fixed.
2. Use a sterile toothpick and obtain a sample from below the gum line in your mouth. (Do not draw blood!). Mix the sample into a drop of water or serum on a slide, then perform a capsule stain on it.
3. Observe using the oil immersion lens. Record your observations of cell morphology and arrangement, cell dimensions, and presence or absence of a capsule in the table on page 58.
4. Dispose of the specimen slides in a jar of disinfectant after use.

REFERENCES

Murray, R. G. E., Raymond N. Doetsch, and C. F. Robinow. 1994. Page 35 in *Methods for General and Molecular Bacteriology*, edited by Philipp Gerhardt, R. G. E. Murray, Willis A. Wood, and Noel R. Krieg. American Society for Microbiology, Washington, DC.

Norris, J. R. and Helen Swain. 1971. Chapter II in *Methods in Microbiology*, Volume 5A, edited by J. R. Norris and D. W. Ribbons. Academic Press, Ltd, London.

■ Figure 3-14 Procedural Diagram—Capsule Stain

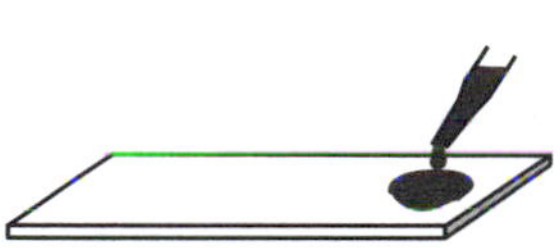

1. Begin with a drop of Congo Red stain at one end of a clean slide. Add a drop of serum.

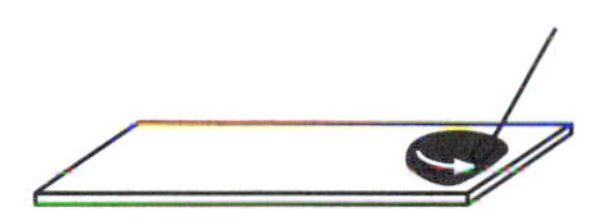

2. Aseptically add organisms and emulsify with a loop. Do not over-inoculate and avoid spattering the mixture. Sterilize the loop after emulsifying.

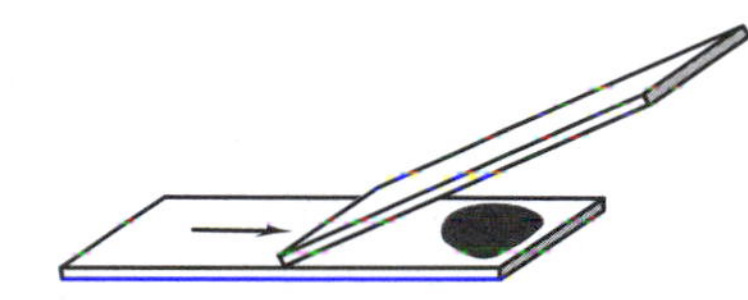

3. Take a second clean slide, place it on the surface of the first slide, and draw it back into the drop.

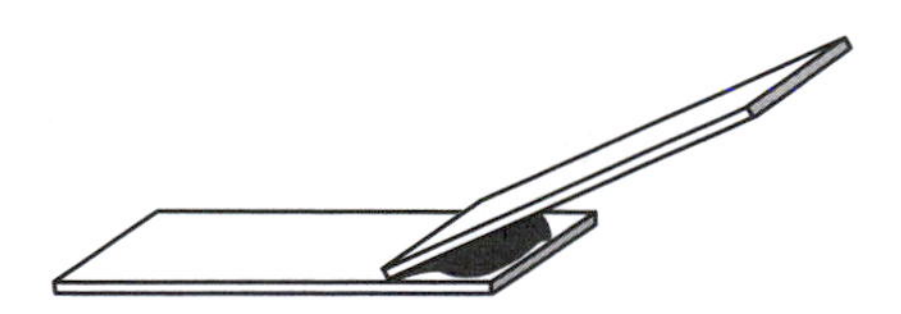

4. When the drop flows across the width of the spreader slide...

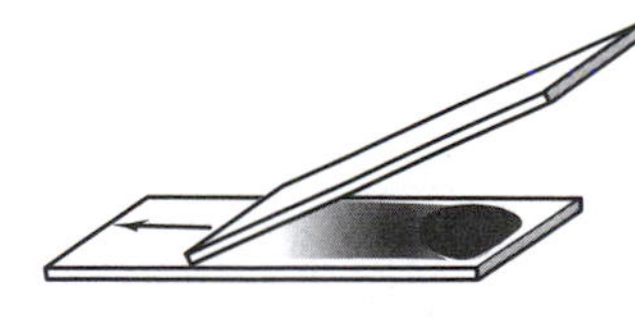

5. ...push the spreader slide to the other end. Dispose of the spreader slide in a jar of disinfectant.

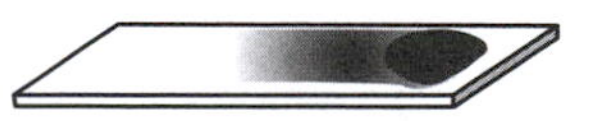

6. Air dry and do NOT heat fix.

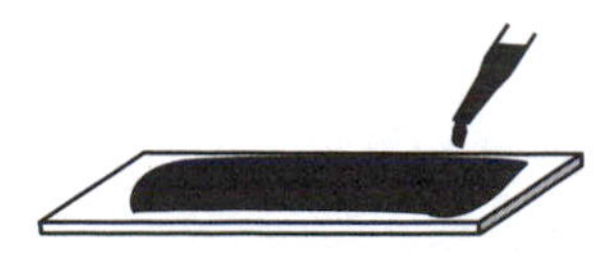

7. Flood the slide with Maneval's Stain for 1 minute. Rinse with water.

8. Gently blot dry in a tablet of bibulous paper. Do not rub. Observe under oil immersion.

■ Observations and Interpretations

Record your observations in the table.

Observations and Interpretations				
Organism	Cellular Morphology and Arrangement (Include a sketch)	Cell Dimensions	Capsule (+/-)	Width of Capsule, If Present

3-9 Endospore Stain

Photographic Atlas Reference
Spore Stain Page 41

Materials Needed For This Exercise

Per Student

- Clean glass microscope slides
- Malachite green stain
- Safranin stain
- Heating apparatus (steam apparatus or hot plate)
- Bibulous paper
- Staining tray
- Staining screen
- Slide holder
- Disposable gloves
- Lab coat or apron
- Goggles
- Nonsterile Petri dish for transporting slides
- Recommended organisms:
  18 to 24 hour Nutrient Agar slant pure culture of *Staphylococcus epidermidis*
  48 hour and 5 day Nutrient Agar slant pure cultures of *Bacillus cereus*

Procedure

1. Prepare and heat-fix a smear of each culture on the same slide as illustrated in Figure 3-7.
2. Follow the instructions in the Procedural Diagram in Figure 3-15. Use a steaming apparatus like the one shown in Figure 3-12. Be sure to use with adequate ventilation and eye protection. If the slide must be carried to and from the steaming apparatus, put it in a covered Petri dish.
3. Observe using the oil immersion lens. Record your observations of cell morphology and arrangement, cell dimensions, and spore presence, position and shape in the table provided below.
4. Dispose of specimen slides in a disinfectant jar.

References

Claus, G. William. 1989. Chapter 9 in *Understanding Microbes—A Laboratory Textbook for Microbiology*. W. H. Freeman and Company, New York, NY.

Murray, R. G. E., Raymond N. Doetsch, and C. F. Robinow. 1994. Page 34 in *Methods for General and Molecular Bacteriology*, edited by Philipp Gerhardt, R. G. E. Murray, Willis A. Wood, and Noel R. Krieg. American Society for Microbiology, Washington, DC.

■ Figure 3-15 Procedural Diagram—Schaeffer-Fulton Spore Stain
Steam the staining preparation; do not boil it. Be sure to perform this procedure with adequate ventilation (preferably a fume hood) and eye protection.

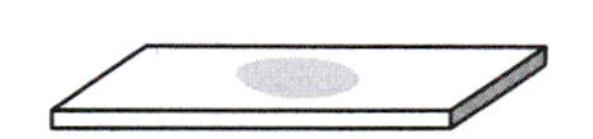

1. Begin with a heat-fixed emulsion.

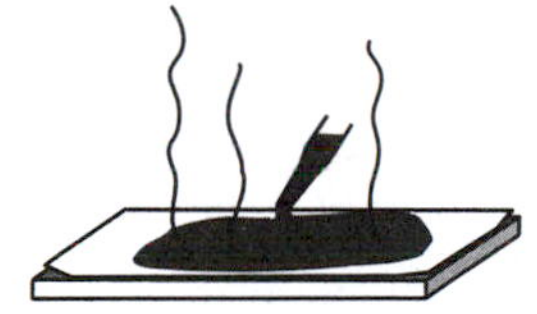

2. Cover the smear with a strip of bibulous paper.
Apply Malachite Green stain.
Steam (as shown in Figure 3-12) for 10 minutes.
Keep the paper moist with stain.
Perform this step with adequate ventillation and eye protection.

3. Grasp the slide with a slide holder.
Remove the paper and dispose of it properly.
Gently rinse the slide with distilled water.

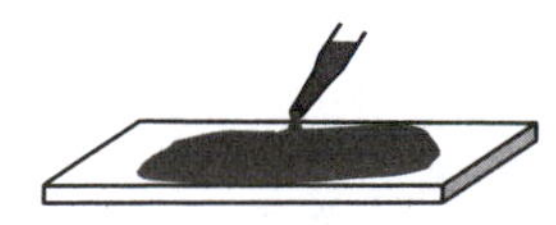

4. Counterstain with Safranin stain for 1 minute.
Rinse with water.

5. Gently blot dry in a tablet of bibulous paper.
Do not rub.
Observe under oil immersion.

Observations and Interpretations

Record your observations in the table.

Observations and Interpretations					
Organism	Cellular Morphology and Arrangement (Include a sketch)	Cell Dimensions	Spores (+/-)	Spore Shape	Spore Position

3-10 WET MOUNT AND HANGING DROP PREPARATIONS

Photographic Atlas Reference
Wet Mount and Hanging Drop Preparations
Page 43

Materials Needed For This Exercise

Per Student

- Depression slide and cover glass
- Clean microscope slides and cover glasses
- Petroleum jelly
- Toothpick
- Recommended organisms (overnight cultures grown on solid media):
  - *Proteus vulgaris*
  - *Aeromonas hydrophila*
  - *Staphylococcus epidermidis*

Procedure—Hanging Drop Preparation (for long-term observation)

1. Follow the procedure illustrated in Figure 3-16 for each specimen.
2. Observe under high dry or oil immersion and record your results in the table below.
3. When finished, remove the cover glass from the slide with an inoculating loop and soak both in a disinfectant jar for at least 15 minutes. Flame the loop. Rinse the slide with 95% ethanol to remove the petroleum jelly, then with water to remove the alcohol. Dry the slide for reuse.

Procedure—Wet Mount Preparation (for short-term observation)

1. Place a drop of water on a clean glass slide (Figure 3-6).
2. Add bacteria to the drop. Don't over-inoculate. Flame the loop after transfer.
3. Gently lower a cover glass with your loop supporting one side over the drop of water. Avoid trapping air bubbles.
4. Observe under high dry or oil immersion and record your results in the table provided.
5. Dispose of the slide and cover glass in a disinfectant jar.

References

Iino, Tetsuo and Masatoshi Enomoto. 1969. Chapter IV in *Methods in Microbiology*, Volume 1, edited by J. R. Norris and D. W. Ribbins. Academic Press, Ltd., London.

Murray, R. G. E., Raymond N. Doetsch, and C. F. Robinow. 1994. Page 26 in *Methods for General and Molecular Bacteriology*, edited by Philipp Gerhardt, R. G. E. Murray, Willis A. Wood, and Noel R. Krieg. American Society for Microbiology, Washington, DC.

Quesnel, Louis B. 1969. Chapter X in *Methods in Microbiology*, Volume 1, edited by J. R. Norris and D. W. Ribbins. Academic Press, Ltd., London.

■ Figure 3-16 Procedural Diagram–The Hanging Drop Preparation
The hanging drop method is used for long-term observation of a living specimen.

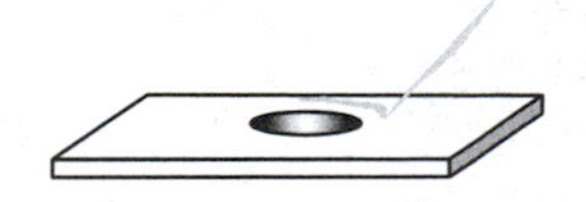

1. Apply a light ring of petroleum jelly around the well of a depression slide with a toothpick.

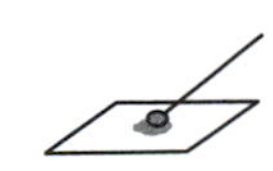

2. Apply a drop of water to a cover glass. Do not use too much water.

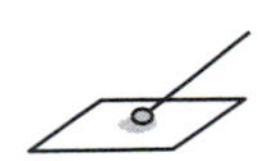

3. Aseptically add a drop of bacteria to the water. Flame your loop after the transfer.

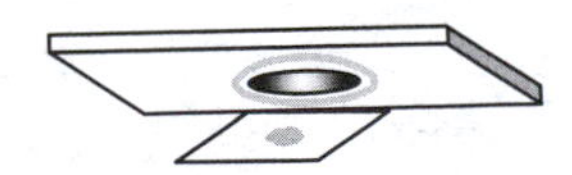

4. Invert the depression slide so the drop is centered in the well. Gently press until the petroleum jelly has created a seal between the slide and cover glass.

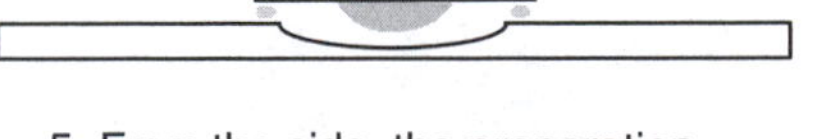

5. From the side, the preparation should look like this. Notice that the drop is "hanging", and is not in contact with the depression slide. Observe under high dry or oil immersion.

■ Observations and Interpretations

Record your observations in the table.

OBSERVATIONS AND INTERPRETATIONS				
ORGANISM	PROCEDURE (WET MOUNT OR HANGING DROP)	CELLULAR MORPHOLOGY AND ARRANGEMENT (INCLUDE A SKETCH)	CELL DIMENSIONS	MOTILITY (+/-)

3-11 Morphological Unknown

Introduction

In this exercise you will be given one pure bacterial culture selected from the organisms listed in Table 3-4. Your job will be to identify it using only the staining techniques covered in Section Three.

Your first task is to convert the information organized in Table 3-4 into flow chart form. A flow chart is simply a visual tool to illustrate the process of elimination that is the foundation of unknown determination. We have started it for you in Appendix E by giving you a few of the branches and listing appropriate organisms. You must complete it by adding necessary branches until you have shown a path to identify each of the organisms. The Procedure contains a detailed explanation of the process.

Once you have designed your flow chart, you will run one stain at a time on your organism. As you match your staining results with those in the flow chart, you will follow a path to identification of your unknown.

A final differential stain will be performed as a **confirmatory test**. It will serve as further evidence that you have correctly identified your unknown.

Materials Needed For This Exercise

Per Student

- Microscope
- Clean microscope slides
- Clean cover glasses
- Gram stain kit
- Acid fast stain kit
- Capsule stain kit
- Spore stain kit
- Bunsen burner
- Striker
- Inoculating loop
- Unknown organisms in numbered tubes (one per student). Fresh slant cultures of*
  - *Bacillus coagulans*
  - *Citrobacter diversus*
  - *Corynebacterium xerosis*
  - *Flavobacterium capsulatum*
  - *Micrococcus roseus*
  - *Mycobacterium smegmatis* or *Mycobacterium phlei*
  - *Neisseria sicca* or *Moraxella catarrhalis*
  - *Rhodospirillum rubrum*
  - *Shigella flexneri*
  - *Staphylococcus epidermidis*
  - *Lactococcus lactis*
- Gram-positive and Gram-negative control organisms (one set per group)—18–24 hour Trypticase Soy Agar slant cultures in tubes labeled with Gram reaction.
- Sterile Trypticase Soy Broth tubes (one per student)
- Sterile Trypticase Soy Agar slants (one per student)

* Cultures should have abundant growth.

Procedure

1. Using the information contained in Table 3-4, complete the flow chart in Appendix E. There are eleven organisms and each should occupy a solitary position at the end of a branch. Do not include stains in a branch that do not differentiate between organisms.
2. Obtain one unknown slant culture. Record its number in the box labeled "Unknown #" in the table provided.
3. Perform a Gram stain of the organism. The stain should be run with known Gram-positive and Gram-negative controls to verify your technique (see Figure 5-3 in the *Photographic Atlas*). The Gram stain will not only provide information on Gram reaction, but will also allow observation of cell size, shape, and arrangement. Enter all these and the date in the appropriate boxes of the table provided. (**Note**: cell size for the unknown candidates is not given and should not be used to differentiate the species. To give you an idea of typical cell sizes, the cocci should be about 1 μm in diameter and the rods 1 to 5 μm in length and about 1 μm in width.)
4. If you have cocci, or are unclear about cell arrangement, aseptically transfer your unknown to sterile Trypticase Soy broth and examine it again in 24–48 hours using a crystal violet or carbolfuchsin simple stain. Cell arrangement is often easier to interpret using cells grown in broth.
5. Based on your results, follow the appropriate branches in the flow chart until you reach the list of possible organisms that matches your unknown's Gram reaction and cell morphology.
6. Perform the next stain and determine to which new branch of the flow chart your unknown belongs. Record the result and date of this stain in the table provided.
7. Repeat Step 6 until you eliminate all but one organism in the flow chart—this is your unknown! (If you are not doing all the stains on the day the unknowns were handed out, be sure to inoculate appropriate media so you will have fresh cultures to work with. If you need to do a spore stain, incubate your original culture for another 24–48 hours, then perform the stain.)
8. After you identify your unknown, perform one more stain to confirm your result. This confirmatory test should be one not previously performed and should be added to your flow chart. The result of this test should agree with the predicted result (as given in Table 3-4). If your confirmatory test result doesn't match the expected result, repeat any suspect stains and find the source of error.
9. Use a colored marker to highlight the path on the flow chart that leads to your unknown. Have your instructor check your work.

These results are typical for the species listed. Your particular strains may vary due to their genetics, their age, or the environment in which they are grown.

TABLE 3-4 TABLE OF RESULTS FOR ORGANISMS USED IN THE EXERCISE

Organism	Gram Stain	Cell Morphology	Cell Arrangement (in broth)	Acid-Fast Stain	Motility (Wet Mount)	Capsule	Spore Stain (Run on cultures older than 48 hours)
Bacillus coagulans	+	rod	single cells and short chains	–	+	–	+
Citrobacter diversus	–	rod	single cells or pairs	–	+	–	–
Corynebacterium xerosis	+	rod	single cells or multiples in angular or palisade arrangement	–	–	–	–
Flavobacterium capsulatum	–	rod	single cells	–	–	+	–
Micrococcus roseus	+	coccus	pairs or tetrads	–	–	–	–
Mycobacterium smegmatis	weak +	rod	single cells or branched; sometimes in dense clusters	+	–	–	–
Neisseria sicca	–	coccus	pairs with adjacent sides flattened	–	–	v	–
Rhodospirillum rubrum	–	spirillum or bent rod	single	–	–	–	–
Shigella flexneri	–	rod	single	–	–	–	–
Staphylococcus epidermidis	+	coccus	singles, pairs, tetrads or clusters	–	–	–	–
Lactococcus lactis	+	ovoid cocci (appearing stretched in the direction of the chain or pair)	pairs or chains	–	–	+	–

■ Record your results in the table. Include the date each test was run.

OBSERVATIONS AND INTERPRETATIONS

Unknown #	Gram Stain	Cell Morphology	Cell Arrangement (in broth)	Cell Dimensions (μm)	Acid-Fast Stain	Motility (Wet Mount)	Capsule	Spore Stain
Date Run								
Result								

SECTION FOUR

Selective Media

4

Individual microbial species in a **mixed culture** must be isolated and cultivated as **pure cultures** before they can be correctly identified or tested. Exercises 1-3 and 1-4 demonstrate the most common means of separating different organisms in a mixed culture. The purpose of this section is to introduce you to **selective media**—the media most commonly used in microbial isolation procedures.

Selective media are designed to encourage growth of certain types of organisms while inhibiting growth of others. The media considered in this section are plated media used specifically to isolate pathogenic Gram-negative bacilli or Gram-positive cocci from human or environmental samples containing a mixture of organisms. Some selective media contain indicators that expose differences between organisms. Such media are considered to be selective and **differential**. Refer to Section Five for more information on differential media.

Clinical microbiologists, familiar with human pathogens and the types of infections they cause, choose selective media that will screen out normal flora also likely to be in the sample. Environmental microbiologists often choose selective and differential media to detect **coliform** bacteria. Coliforms are common inhabitants of the human intestinal tract. As such, their presence in the environment is strong evidence of fecal contamination.

In the exercises that follow, you will examine some commonly used selective (and often differential) media for the isolation of Gram-positive cocci and Gram-negative rods. For ease of instruction and to conserve media the procedures call for "spot" inoculations with several pure cultures on a single plate (see Appendix B). In a typical clinical situation, this media would be streaked for isolation since virtually all specimens contain a mixture of organisms. A spot inoculation is simply a 1–2 centimeter streak made on the agar surface and enables observation and comparison of individual reactions.

All of the exercises contain a Table of Results identifying the various reactions likely produced on a particular medium. Each table includes color results, interpretations of results, symbols used to quickly identify the various reactions, and presumptive identification of typical organisms encountered with these media.

It should be understood that presumptive identification is not final identification. It is an "educated guess" based on evidence provided by the selective/differential medium coupled with information about the origin of a sample.

Selective Media for Isolation of Gram-Positive Cocci

Gram-negative organisms frequently inhibit growth of Gram-positives when cultivated together. Therefore, when looking for staphylococci or streptococci in a clinical sample, it may be necessary to begin by streaking the unknown mixture onto a selective medium that inhibits Gram-negative growth.

The selective media introduced in this unit are used to isolate (and sometimes presumptively identify) streptococci and staphylococci in human samples. Phenylethyl Alcohol Agar is a simple selective medium designed to inhibit Gram-negative organisms. Mannitol Salt Agar (selective *and* differential) was developed to favor growth of *Staphylococcus* and to differentiate pathogenic from nonpathogenic members of the genus.

4-1 MANNITOL SALT AGAR

Photographic Atlas Reference
Mannitol Salt Agar Page 18

Materials Needed For This Exercise

Per Student Group

- One Mannitol Salt Agar plate
- One Nutrient Agar plate
- Fresh broth cultures of:
  - *Staphylococcus aureus*
  - *Staphylococcus epidermidis*
  - *Escherichia coli*
  - *Enterobacter aerogenes*

Procedure

Lab One

1. Mix each culture well.
2. Using a permanent marker, divide the bottom of each plate into four sectors.
3. Label the plates with the organisms' names, your name, and the date.
4. Spot inoculate the sectors on the Mannitol Salt Agar plate with the test organisms.
5. Repeat step 4 with the Nutrient Agar plate.
6. Invert the plates and incubate at 35°C for 24 to 48 hours.

Lab Two

1. Examine and compare the plates for color and quality of growth.
2. Record your results in the space provided.

References

Baron, Ellen Jo, Lance R. Peterson, and Sydney M. Finegold. 1994. Chapter 25 in *Bailey and Scott's Diagnostic Microbiology*, 9th Ed. Mosby-Yearbook, St. Louis, MO.

Delost, Maria Danessa. 1997. Page 112 in *Introduction to Diagnostic Microbiology, a Text and Workbook*. Mosby, Inc., St. Louis, MO.

DIFCO Laboratories. 1984. Page 558 in *DIFCO Manual*, 10th Ed. DIFCO Laboratories, Detroit, MI.

Power, David A. and Peggy J. McCuen. 1988. Page 193 in *Manual of BBL® Products and Laboratory Procedures*, 6th Ed. Becton Dickinson Microbiology Systems, Cockeysville, MD.

TABLE 4-1 MANNITOL SALT AGAR RESULTS AND INTERPRETATIONS

Result	Interpretation	Presumptive ID
Poor growth or no growth (P)	Organism is inhibited by NaCl	Not *Staphylococcus*
Good growth (G)	Organism is not inhibited by NaCl	*Staphylococcus*
Yellow growth or halo (Y)	Organism produces acid (A) from mannitol fermentation	Possible pathogenic *Staphylococcus aureus*
Red growth (no halo) (R)	Organism does not ferment mannitol. No reaction (NR)	Nonpathogenic *Staphylococcus*

■ OBSERVATIONS AND INTERPRETATIONS

Refer to Table 4-1 when recording your results and interpretations in the table below. Use abbreviations or symbols as needed.

OBSERVATIONS AND INTERPRETATIONS				
ORGANISM	GROWTH (P/G)		MSA COLOR (Y/R)	INTERPRETATION
	MSA	NA		

4-2 PHENYLETHYL ALCOHOL AGAR

Photographic Atlas Reference
Phenylethyl Alcohol Agar Page 19

MATERIALS NEEDED FOR THIS EXERCISE

Per Student Group

- One PEA plate
- One Nutrient Agar Plate
- Fresh broth cultures of:
  - *Escherichia coli*
  - *Enterococcus faecalis*
  - *Staphylococcus aureus*

PROCEDURE

Lab One

1. Mix each culture well.
2. Using a permanent marker, divide the bottom of each plate into three sectors.
3. Label the plates with the organisms' names, your name, and the date.
4. Spot inoculate the sectors on the PEA plate with the test organisms.
5. Repeat step 4 with the Nutrient Agar plate.
6. Invert the plates and incubate at 35°C for 24 to 48 hours.

Lab Two

1. Examine and compare the plates for color and quality of growth.
2. Record your results in the space provided.

REFERENCES

DIFCO Laboratories. 1984. Page 667 in *DIFCO Manual*, 10th Ed. DIFCO Laboratories, Detroit, MI.

Power, David A. and Peggy J. McCuen. 1988. Page 223 in *Manual of BBL® Products and Laboratory Procedures*, 6th Ed. Becton Dickinson Microbiology Systems, Cockeysville, MD.

TABLE 4-2 PEA RESULTS AND INTERPRETATIONS

RESULT	INTERPRETATION	PRESUMPTIVE ID
Poor growth or no growth (P)	Organism is inhibited by phenylethyl alcohol	Probable Gram-negative organism
Good growth (G)	Organism is not inhibited by phenylethyl alcohol	Probable *Staphylococcus*, *Streptococcus*, *Enterococcus*, or *Lactococcus*

■ OBSERVATIONS AND INTERPRETATIONS

Refer to Table 4-2 when recording your results and interpretations in the table below.

OBSERVATIONS AND INTERPRETATIONS

ORGANISM	GROWTH (P/G)		INTERPRETATION
	PEA	NA	

Selective Media for Isolation of Gram-Negative Rods

Most Gram-negative rods found in clinical samples are members of *Enterobacteriaceae*—the enteric "gut" bacteria. The media used to isolate and differentiate these organisms from each other (and other Gram-negative pathogens) frequently combine several selective and differential elements in order to gather as much information as possible with a single inoculation and incubation.

Two examples are Hektoen Enteric (HE) Agar and Xylose Lysine Desoxycholate (XLD) Agar. HE Agar differentiates *Salmonella* and *Shigella* both from each other and from other enterics based on ability to overcome the inhibitory effects of bile, reduce sulfer to H_2S, and ferment lactose, sucrose or saliein. XLD Agar favors growth of *Salmonella, Shigella,* or *Providencia* based on their ability to overcome the inhibitory effects of desoxycholate, and differentiates them based on their ability to reduce sulfur to H_2S, decarboxylate the amino acid lysine, and ferment xylose or lactose.

Other media such as Eosin Methylene Blue Agar and Endo Agar test for the presence of coliforms in environmental samples. Loosely defined, coliforms are enteric organisms that produce acid and gas from the fermentation of lactose. Their presence in the environment suggests fecal contamination and the possible presence of more serious pathogens.

Other media included in this unit are Desoxycholate Agar and MacConkey Agar. Each is selective for Gram-negative organisms and differentiates based on lactose fermentation. Refer to Section 5 in the *Photographic Atlas* for more information on differential tests.

4-3 Desoxycholate Agar

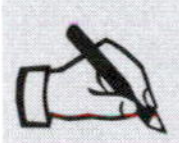

Photographic Atlas Reference
Desoxycholate Agar Page 14

Materials Needed For This Exercise

Per Student Group

- One DOC plate
- One Nutrient Agar plate
- Fresh broth cultures of:
  - *Staphylococcus aureus*
  - *Salmonella typhimurium*
  - *Escherichia coli*

Procedure

Lab One

1. Mix each culture well.
2. Using a permanent marker, divide the bottom of each plate into three sectors.
3. Label the plates with the organisms' names, your name, and the date.
4. Spot inoculate the sectors on the DOC plate with the test organisms.
5. Repeat step 4 with the nutrient agar plate.
6. Invert the plates and incubate at 35°C for 24 to 48 hours.

Lab Two

1. Examine and compare the plates for color and quality of growth.
2. Record your results in the space provided.

Table 4-3 DOC Results and Interpretations

Result	Interpretation	Presumptive ID
Poor growth or no growth (P)	Organism is inhibited by desoxycholate	Gram-positive
Good growth (G)	Organism is not inhibited by desoxycholate	Gram-negative
Growth is red or pink (R)	Organism ferments lactose to acid (A) end products	Possible coliform
Growth is colorless (not red or pink) (C)	Organism does not ferment lactose. No reaction (NR)	Noncoliform

■ Observations and Interpretations

Refer to Table 4-3 when recording your results and interpretations in the table below. Use abbreviations or symbols as needed.

OBSERVATIONS AND INTERPRETATIONS				
Organism	**Growth (P/G)**		**DOC Growth Color (R/C)**	**Interpretation**
	DOC	**NA**		

References

DIFCO Laboratories. 1984. Page 274 in *DIFCO Manual*, 10th Ed. DIFCO Laboratories, Detroit, MI.

Koneman, Elmer W., Stephen D. Allen, William M. Janda, Paul C. Schreckenberger, and Washington C. Winn, Jr. 1997. Chapters 4 and 6 in *Color Atlas and Textbook of Diagnostic Microbiology*, 5th Ed. J. B. Lippincott Company, Philadelphia, PA.

Power, David A. and Peggy J. McCuen. 1988. Page 144 in *Manual of BBL® Products and Laboratory Procedures*, 6th Ed. Becton Dickinson Microbiology Systems, Cockeysville, MD.

4-4 ENDO AGAR

Photographic Atlas Reference
Endo Agar Page 16

MATERIALS NEEDED FOR THIS EXERCISE

Per Student Group

- One Endo Agar plate
- One Nutrient Agar plate
- Fresh broth cultures of:
  - *Escherichia coli*
  - *Proteus mirabilis*
  - *Micrococcus luteus*

PROCEDURE

Lab One

1. Mix each culture well.
2. Using a permanent marker, divide the bottom of each plate into three sectors.
3. Label the plates with the organisms' names, your name, and the date.
4. Spot inoculate the sectors on the Endo Agar plate with the test organisms.
5. Repeat step 4 with the Nutrient Agar plate.
6. Invert the plates and incubate at 35°C for 24 to 48 hours.

Lab Two

1. Examine and compare the plates for color and quality of growth.
2. Record your results in the space provided.

REFERENCES

DIFCO Laboratories. 1984. Page 324 in *DIFCO Manual*, 10th Ed. DIFCO Laboratories, Detroit, MI.

Power, David A. and Peggy J. McCuen. 1988. Page 153 in *Manual of BBL® Products and Laboratory Procedures*, 6th Ed. Becton Dickinson Microbiology Systems, Cockeysville, MD.

TABLE 4-4 ENDO AGAR RESULTS AND INTERPRETATIONS

RESULT	INTERPRETATION	PRESUMPTIVE ID
Poor growth or no growth (P)	Organism is inhibited by sodium sulfite or basic fuchsin	Gram-positive
Good growth (G)	Organism is not inhibited by sodium sulfite or basic fuchsin	Gram-negative
Red or pink growth (R)	Organism ferments lactose to acid end products	Possible coliform
Red or pink growth with darkened medium and/or metallic sheen (RD)	Organism ferments lactose rapidly and produces large amount of acid (A)	Probable coliform
Growth is "colorless" (not red or pink and does not produce metallic sheen) (C)	Organism does not ferment lactose. No reaction (NR)	Noncoliform

■ OBSERVATIONS AND INTERPRETATIONS

Refer to Table 4-4 when recording and interpreting your results in the table below. Use abbreviations or symbols as needed.

OBSERVATIONS AND INTERPRETATIONS

ORGANISM	GROWTH (G/P)		ENDO GROWTH COLOR (R/RD/C)	INTERPRETATION
	ENDO	NA		

4-5 EOSIN METHYLENE BLUE AGAR

Photographic Atlas Reference
Eosin Methylene Blue Agar Page 16

MATERIALS NEEDED FOR THIS EXERCISE

Per Student Group

- One EMB plate
- One Nutrient Agar plate
- Fresh broth cultures of:
  - *Staphylococcus aureus*
  - *Escherichia coli*
  - *Proteus vulgaris*

PROCEDURE

Lab One

1. Mix each culture well.
2. Using a permanent marker, divide the bottom of each plate into three sectors.
3. Label the plates with the organisms' names, your name, and the date.
4. Spot inoculate the three sectors on the EMB plate with the test organisms.
5. Repeat step 4 with the Nutrient Agar plate.
6. Invert the plates and incubate at 35°C for 24 to 48 hours.

Lab Two

1. Examine and compare the plates for color and quality of growth.
2. Record your results in the space provided.

REFERENCES

DIFCO Laboratories. 1984. Page 324 in *DIFCO Manual*, 10th Ed. DIFCO Laboratories, Detroit, MI.

Koneman, Elmer W., Stephen D. Allen, William M. Janda, Paul C. Schreckenberger, and Washington C. Winn, Jr. 1997. Chapter 4 in *Color Atlas and Textbook of Diagnostic Microbiology*, 5th Ed. J. B. Lippincott Company, Philadelphia, PA.

Power, David A. and Peggy J. McCuen. 1988. Page 153 in *Manual of BBL® Products and Laboratory Procedures*, 6th Ed. Becton Dickinson Microbiology Systems, Cockeysville, MD.

TABLE 4-5 EMB RESULTS AND INTERPRETATIONS

RESULT	INTERPRETATION	PRESUMPTIVE ID
Poor growth or no growth (P)	Organism is inhibited by eosin and methylene blue	Gram-positive
Good growth (G)	Organism is not inhibited by eosin and methylene blue	Gram-negative
Growth is pink and mucoid (Pi)	Organism ferments lactose with little acid production (a)	Possible coliform
Growth is "dark" (purple to black, with or without green metallic sheen) (D)	Organism ferments lactose and/or sucrose with much acid production (A)	Probable coliform
Growth is "colorless" (no pink, purple, or metallic sheen)(C)	Organism does not ferment lactose or sucrose. No reaction (NR)	Noncoliform

■ OBSERVATIONS AND INTERPRETATIONS

Refer to Table 4-5 when recording and interpreting your results in the table below. Use abbreviations or symbols as needed.

OBSERVATIONS AND INTERPRETATIONS

ORGANISM	GROWTH (P/G)		EMB GROWTH COLOR (PI/D/C)	INTERPRETATION
	EMB	NA		

4-6 MacConkey Agar

Photographic Atlas Reference
MacConkey Agar Page 18

Materials Needed For This Exercise

Per Student Group

- One MacConkey Agar plate
- One Nutrient Agar plate
- Fresh broth cultures of:
  - *Staphylococcus aureus*
  - *Enterobacter aerogenes*
  - *Escherichia coli*
  - *Proteus vulgaris*

Procedure

Lab One

1. Mix each culture well.
2. Using a permanent marker, divide the bottom of each plate into four sectors.
3. Label the plates with the organisms' names, your name, and the date.
4. Spot inoculate the sectors on the MacConkey Agar plate with the test organisms.
5. Repeat step 4 with the Nutrient Agar plate.
6. Invert the plates and incubate at 35°C for 24 to 48 hours.

Lab Two

1. Examine and compare the plates for color and quality of growth.
2. Record your results in the space provided.

References

Baron, Ellen Jo, Lance R. Peterson and Sydney M. Finegold. 1994. Chapters 9 and 28 in *Bailey and Scott's Diagnostic Microbiology*, 9th Ed. Mosby-Yearbook, St. Louis, MO.

DIFCO Laboratories. 1984. Page 546 in *DIFCO Manual*, 10th Ed. DIFCO Laboratories, Detroit, MI.

Forbes, Betty A., Daniel F. Sahm, and Alice S. Weissfeld. 2002. Page 139 in *Bailey and Scott's Diagnostic Microbiology*, 11th Ed. Mosby-Yearbook, St. Louis, MO.

Koneman, Elmer W., Stephen D. Allen, William M. Janda, Paul C. Schreckenberger, and Washington C. Winn, Jr. 1997. Chapter 4 in *Color Atlas and Textbook of Diagnostic Microbiology*, 5th Ed. J. B. Lippincott Company, Philadelphia, PA.

Power, David A. and Peggy J. McCuen. 1988. Page 189 in *Manual of BBL® Products and Laboratory Procedures*, 6th Ed. Becton Dickinson Microbiology Systems, Cockeysville, MD.

Table 4-6 MacConkey Agar Results and Interpretations

Result	Interpretation	Presumptive ID
Poor growth or no growth (P)	Organism is inhibited by crystal violet and/or bile	Gram-positive
Good growth (G)	Organism is not inhibited by crystal violet or bile	Gram-negative
Pink to red growth with or without bile precipitate (R)	Organism produces acid from lactose fermentation (A)	Probable coliform
Growth is "colorless" (not red or pink) (C)	Organism does not ferment lactose. No reaction (NR)	Noncoliform

■ Observations and Interpretations

Refer to Table 4-6 when recording your results and interpretations in the table below. Use abbreviations or symbols as needed.

Observations and Interpretations

Organism	Growth (P/G)		Mac Growth Color (R/C)	Interpretation
	MAC	NA		

4-7 HEKTOEN ENTERIC AGAR

Photographic Atlas Reference
Hektoen Enteric Agar Page 17

MATERIALS NEEDED FOR THIS EXERCISE

Per Student Group

- One Hektoen Enteric Agar plate
- One Nutrient Agar plate
- Fresh cultures of:
  - *Escherichia coli*
  - *Shigella flexneri*
  - *Salmonella typhimurium*
  - *Staphylococcus aureus*

PROCEDURE

Lab One

1. Mix each culture well.
2. Using a permanent marker, divide the bottom of each plate into four sectors.
3. Label the plates with the organisms' names, your name, and the date.
4. Spot inoculate the four sectors on the HE plate with the test organisms.
5. Repeat step 4 with the Nutrient Agar plate.
6. Invert the plates and incubate aerobically at 35°C for 48 hours.

Lab Two

1. Examine and compare the plates for color and quality of growth.
2. Record your results in the space provided.

REFERENCES

Baron, Ellen Jo, Lance R. Peterson, and Sydney M. Finegold. 1994. Chapter 9 in *Bailey and Scott's Diagnostic Microbiology*, 9th Ed. Mosby-Yearbook, St. Louis, MO.

DIFCO Laboratories. 1984. Page 455 in *DIFCO Manual*, 10th Ed. DIFCO Laboratories, Detroit, MI.

Forbes, Betty A., Daniel F. Sahm, and Alice S. Weissfeld. 2002. Page 139 in *Bailey and Scott's Diagnostic Microbiology*, 11th Ed. Mosby-Yearbook, St. Louis, MO.

Koneman, Elmer W., Stephen D. Allen, William M. Janda, Paul C. Schreckenberger, and Washington C. Winn, Jr. 1997. Chapter 4 in *Color Atlas and Textbook of Diagnostic Microbiology*, 5th Ed. J. B. Lippincott Company, Philadelphia, PA.

Power, David A. and Peggy J. McCuen. 1988. Page 167 in *Manual of BBL® Products and Laboratory Procedures*, 6th Ed. Becton Dickinson Microbiology Systems, Cockeysville, MD.

4-7 HEKTOEN ENTERIC AGAR RESULTS AND INTERPRETATIONS

RESULT	INTERPRETATION	PRESUMPTIVE ID
Poor growth or no growth (P)	Organism is inhibited by bile and/or one of the dyes included	Gram-positive
Good growth (G)	Organism is not inhibited by bile or any of the dyes included	Gram-negative
Pink to orange growth (Pi)	Organism produces acid from lactose fermentation (A)	Not *Shigella* or *Salmonella*
Blue-green growth with black precipitate (Bppt)	Organism does not ferment lactose, but reduces sulfur to hydrogen sulfide (H_2S)	Possible *Salmonella*
Blue-green growth without black precipitate (B)	Organism does not ferment lactose or reduce sulfur. No reaction (NR)	Possible *Shigella* or *Salmonella*

OBSERVATIONS AND INTERPRETATIONS

Refer to Table 4-7 when recording your results and interpretations in the table below. Use abbreviations or symbols as needed.

OBSERVATIONS AND INTERPRETATIONS

ORGANISM	GROWTH (P/G)		HE GROWTH COLOR (PI/BPPT/B)	INTERPRETATION
	HE	NA		

4-8 XYLOSE LYSINE DESOXYCHOLATE AGAR

Photographic Atlas Reference
Xylose Lysine Desoxycholate Agar Page 22

MATERIALS NEEDED FOR THIS EXERCISE

Per Student Group

- One XLD Agar plate
- One Nutrient Agar plate
- Fresh broth cultures of:
    - *Staphylococcus aureus*
    - *Providencia stuartii*
    - *Salmonella typhimurium*
    - *Enterobacter aerogenes*
    - *Proteus mirabilis*

PROCEDURE

Lab One

1. Mix each culture well.
2. Using a permanent marker, divide the bottom of each plate into five sectors.
3. Label the plates with the organisms' names, your name, and the date.
4. Spot inoculate the sectors on the XLD Agar plate with the test organisms.
5. Repeat step 4 with the Nutrient Agar plate.
6. Invert the plates and incubate at 35°C for 18 to 24 hours. (Be sure to remove the plates from the incubator at 24 hours. Reversions from fermentation to decarboxylation [due to exhaustion of sugar in the medium] with the resulting yellow-to-red color shift may lead to false identification as *Shigella* or *Providencia*.)

Lab Two

1. Examine and compare the plates for color and quality of growth.
2. Record your results in the space provided.

REFERENCES

Baron, Ellen Jo, Lance R. Peterson and Sydney M. Finegold. 1994. Chapter 9 in *Bailey and Scott's Diagnostic Microbiology*, 9th Ed. Mosby-Yearbook, St. Louis, MO.

DIFCO Laboratories. 1984. Page 1128 in *DIFCO Manual*, 10th Ed. DIFCO Laboratories, Detroit, MI.

Forbes, Betty A., Daniel F. Sahm, and Alice S. Weissfeld. 2002. Page 140 in *Bailey and Scott's Diagnostic Microbiology*, 11th Ed. Mosby-Yearbook, St. Louis, MO.

Power, David A. and Peggy J. McCuen. 1988. Page 288 in *Manual of BBL® Products and Laboratory Procedures*, 6th Ed. Becton Dickinson Microbiology Systems, Cockeysville, MD.

TABLE 4-8 XLD RESULTS AND INTERPRETATIONS

RESULT	INTERPRETATION	PRESUMPTIVE ID
Poor growth (P)	Organism is inhibited by desoxycholate	Gram-positive
Good growth (G)	Organism is not inhibited by desoxycholate	Gram-negative
Growth is yellow (Y)	Organism produces acid from xylose fermentation (A)	Not *Shigella* or *Providencia*
Red growth with black center (RB)	Organism reduces sulfur to hydrogen sulfide (H_2S)	Not *Shigella* or *Providencia* (probable *Salmonella*)
Red growth without black center (R)	Organism does not ferment xylose or ferments xylose slowly; alkaline products from lysine decarboxylation (K)	Probable *Shigella* or *Providencia*

■ Observations and Interpretations

Refer to Table 4-8 when recording your results and interpretations in the table below. Use abbreviations or symbols as needed.

OBSERVATIONS AND INTERPRETATIONS

Organism	Growth (P/G)		XLD Growth Color (Y/RB/R)	Interpretation
	XLD	NA		

Differential Tests

5

Over most of the last century, microbiologists have differentiated bacteria based on their enormous biochemical diversity. The identification systems used for this are called **differential tests** or **biochemical tests** because they differentiate and sometimes identify microorganisms based on specific biochemical characteristics. As you will see repeatedly in this section, microbial biochemical characteristics are brought about by enzymatic reactions that follow specific and well-understood metabolic pathways. The tests used in this section were selected to illustrate those pathways and means used to identify them.

The first half of the section will introduce you to the tests individually so that you can familiarize yourself with the metabolic pathways and the mechanisms by which specific tests expose them. In the second half of the section you will be given a chance to use some multiple test media designed for rapid identification.

The categories of differential tests included in this section are:

- Energy metabolism including fermentation of carbohydrates and aerobic and anaerobic respiration
- Utilization of a specified medium component
- Decarboxylation and deamination of amino acids
- Hydrolytic reactions requiring intracellular or extracellular enzymes
- Multiple reactions performed in a single combination medium
- Miscellaneous differential tests

As a culmination to your work in this section, you will be given unknown bacteria to identify using these differential tests.

Introduction to Energy Metabolism Tests

Heterotrophic bacteria obtain their energy by means of **respiration** or **fermentation.** Both catabolic systems convert the chemical energy of glucose to high-energy bonds in **adenosine triphosphate (ATP).** In respiration glucose is converted to ATP in three distinct phases—**glycolysis,** the **tricarboxylic acid cycle (Krebs cycle),** and **oxidative phosphorylation** (sometimes called the **electron transport chain** or ETC).

Glycolysis splits the six-carbon glucose molecule into two three-carbon **pyruvic acid molecules** with the production of ATP and reduced **coenzymes.** The Krebs cycle is the complex pathway in which **acetyl-CoA** (from the conversion of pyruvic acid) is oxidized to CO_2 and more coenzymes are reduced. ATP is also a product. The electron transport chain (ETC) is a series of **oxidation-reduction** reactions that receives electrons from the reduced coenzymes produced during glycolysis and the Krebs cycle. At the end of the ETC is an inorganic molecule called the **final electron acceptor (FEA).** When oxygen is the final electron acceptor the respiration is **aerobic.** If the FEA is an inorganic molecule other than oxygen (*e.g.,* sulfate or nitrate) the respiration is **anaerobic.**

In contrast to respiration, fermentation is the metabolic process by which glucose acts as an electron donor and one or more of its organic products act as the final electron acceptor. Reduced carbon compounds in the form of acids, and organic solvents, as well as CO_2 are the typical end products of fermentation.

In this first unit, you will perform a simple test used to determine the ability of an organism to perform oxidative and/or fermentative metabolism of sugars. In the exercises that follow, you will perform tests that demonstrate microbial fermentative characteristics and tests designed to detect specific respiratory constituents or pathways.

5-1 OXIDATION-FERMENTATION TEST

Photographic Atlas Reference
Oxidation-Fermentation Test Page 73

MATERIALS NEEDED FOR THIS EXERCISE

Per Student Group

- Eight OF glucose tubes
- Sterile mineral oil
- Sterile transfer pipettes
- Fresh agar slants of:
  - *Alcaligenes faecalis*
  - *Escherichia coli*
  - *Pseudomonas aeruginosa*

PROCEDURE

Lab One

1. Obtain eight OF tubes. Label six (in three pairs) with the organisms' names, your name, and the date. Label the last pair "control."
2. Stab-inoculate two OF tubes with each test organism. Stab several times to a depth of about one cm from the bottom of the agar. Do not inoculate the controls.
3. Overlay one of each pair of tubes (including the controls) with 3–4 mm of sterile mineral oil (Figure 5-1).
4. Incubate all tubes at 35°C for 48 hours.

Lab Two

1. Examine the tubes for color changes. Re-incubate any OF tubes that have not changed color for an additional 48 hours.
2. Record your results in the table provided.

FIGURE 5-1 Adding the Mineral Oil Layer
Tip the tube slightly to one side and gently add 3–4 mm of sterile mineral oil. Be sure to use a sterile pipette for each tube.

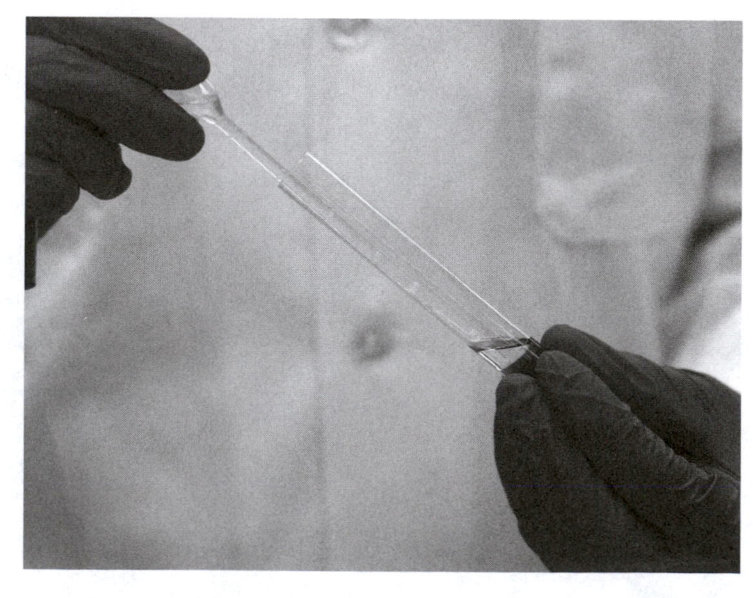

TABLE 5-1 OF TEST RESULTS AND INTERPRETATIONS			
RESULT		**INTERPRETATION**	**SYMBOL**
SEALED	**UNSEALED**		
Green or blue	Yellow at the top	Oxidation	O
Yellow throughout	Yellow throughout	Oxidation and fermentation or fermentation only	O/F
Slightly yellow at the top	Slightly yellow at the top	Oxidation and slow fermentation or slow fermentation only	O/F
Green or blue	Green or blue	No sugar metabolism; Organism is nonsaccharolytic	N

■ OBSERVATIONS AND INTERPRETATIONS

Refer to Table 5-1 when recording your results and interpretations in the table below. (See *Photographic Atlas* Figure 6-78.)

OBSERVATIONS AND INTERPRETATIONS				
ORGANISM	**COLOR RESULTS**		**SYMBOL**	**INTERPRETATION**
	SEALED	**UNSEALED**		
Uninoculated Control				

REFERENCES

Collins, C. H., Patricia M. Lyne, J. M. Grange. 1995. Page 112 in *Collins and Lyne's Microbiological Methods*, 7th Ed. Butterworth-Heinemann, UK.

Delost, Maria Dannessa. 1997. Pages 218–219 in *Introduction to Diagnostic Microbiology.* Mosby, Inc., St. Louis, MO.

DIFCO Laboratories. 1984. Page 625 in *DIFCO Manual*, 10th Ed. DIFCO Laboratories, Detroit, MI.

Forbes, Betty A., Daniel F. Sahm, Alice S. Weissfeld. 2002. Pages 154–155 in *Bailey & Scott's Diagnostic Microbiology*, 11th Ed. Mosby, Inc., St. Louis, MO.

MacFaddin, Jean F. 2000. Page 379 in *Biochemical Tests for Identification of Medical Bacteria*, 3rd Ed. Lippincott Williams & Wilkins, Philadelphia, PA.

Power, David A. and Peggy J. McCuen. 1988. Page 216 in *Manual of BBL® Products and Laboratory Procedures*, 6th Ed. Becton Dickinson Microbiology Systems, Cockeysville, MD.

Smibert, Robert M. and Noel R. Krieg. 1994. Page 625 in *Methods for General and Molecular Bacteriology*, edited by Philipp Gerhardt, R. G. E. Murray, Willis A. Wood, and Noel R. Krieg, American Society for Microbiology, Washington, DC.

Fermentation Tests

Carbohydrate fermentation, as defined at the beginning of this section, is the metabolic process by which an organic molecule acts as an electron donor (so it becomes oxidized) and one or more of its organic products act as the final electron acceptor (FEA). In actuality, the term "carbohydrate fermentation" is rather broadly used to include hydrolysis of disaccharides prior to the fermentation reaction. Thus, a "lactose fermenter" is an organism that splits the disaccharide lactose into glucose and galactose, then ferments the glucose. In addition, the organism may convert the galactose to glucose, which it also ferments. In this section and in supporting material in the *Photographic Atlas* you will frequently see the term "fermenter." Unless it is expressly addressed as otherwise, assume the term to include the initial hydrolysis and/or conversion reactions.

Fermentation of glucose begins with the production of pyruvic acid. Although some organisms use alternate pathways, most bacteria accomplish this by glycolysis (See Appendix A in the *Photographic Atlas*). Pyruvic acid fermentation end products include a variety of organic acids, H_2 or CO_2 gases, and alcohols. The specific end products depend on the specific organism and the substrate fermented (*Photographic Atlas* Table A-6).

In this unit you will perform tests using three differential media formulated to detect fermentation. General-purpose fermentation media typically include one or more carbohydrates and a pH indicator to detect acid formation. Purple Broth and Phenol Red Broth are examples of general-purpose fermentation media. MR-VP broth is a more specialized medium, in that it allows detection of two specific fermentation pathways. The Methyl Red (MR) Test detects bacteria capable of performing a **mixed acid fermentation.** The Voges-Proskauer (VP) Test identifies bacteria able to produce acetoin as part of a **2,3-butanediol fermentation.**

5-2 PHENOL RED BROTH

Photographic Atlas Reference
Phenol Red Broth Page 57

Materials Needed For This Exercise

Per Student Group

- Five PR Glucose Broths with Durham tubes
- Five PR Lactose Broths with Durham tubes
- Five PR Sucrose Broths with Durham tubes
- Five PR Base Broths with Durham tubes
- Fresh cultures of:
  - *Klebsiella pneumoniae*
  - *Shigella flexneri*
  - *Proteus vulgaris*
  - *Alcaligenes faecalis*

Procedure

Lab One

1. Obtain five of each PR broth. Label four of them with the organisms' name, your name, the medium name, and the date. Label one of each "control."
2. Inoculate four base broths and four of each carbohydrate broth with the test organisms. Do not inoculate the control.
3. Incubate all the tubes at 35°C for 24 hours.

Lab Two

1. Using the uninoculated broths for color comparison and Table 5-2 as a guide, examine all tubes and enter your results in the table provided. When indicating the various reactions use the standard symbols as shown with the acid reading first followed by a slash and then the gas reading. K indicates alkalinity and –/– symbolizes no reaction. **Do not try to read these results without control tubes for comparison.**

TABLE 5-2 PR BROTH RESULTS AND INTERPRETATIONS

Result	Interpretation	Symbol
Yellow broth, bubble in tube	Fermentation with acid and gas end products	A/G
Yellow broth, no bubble in tube	Fermentation with acid end products; no gas produced	A/–
Red broth, no bubble in tube	No fermentation	–/–
Pink broth, no bubble in tube	Degradation of peptone; Alkaline end products	K

Observations and Interpretations

Enter your results in the table below using the symbols shown in Table 5-2. (See *Photographic Atlas* Figure 6-38.)

Observations and Interpretations					
Organism	Results				Interpretation
	PR Base	PR Glucose	PR Lactose	PR Sucrose	
Uninoculated Control					

References

DIFCO Laboratories. 1984. Page 660 in *DIFCO Manual*, 10th Ed. DIFCO Laboratories, Detroit, MI.

Lányi, B. 1987. Page 44 in *Methods in Microbiology*, Vol. 19, edited by R. R. Colwell and R. Grigorova, Academic Press Inc., New York.

MacFaddin, Jean F. 2000. Page 57 in *Biochemical Tests for Identification of Medical Bacteria*, 3rd Ed. Lippincott Williams & Wilkins, Philadelphia, PA.

Power, David A. and Peggy J. McCuen. 1988. Page 220 in *Manual of BBL® Products and Laboratory Procedures*, 6th Ed. Becton Dickinson Microbiology Systems, Cockeysville, MD.

5-3 PURPLE BROTH

Photographic Atlas Reference
Purple Broth Page 57

MATERIALS NEEDED FOR THIS EXERCISE

Per Student Group

- Five Purple Glucose Broths with Durham tubes
- Five Purple Lactose Broths with Durham tubes
- Five Purple Sucrose Broths with Durham tubes
- Five Purple Base Broths with Durham tubes
- Fresh cultures of:
  Klebsiella pneumoniae
  Shigella flexneri
  Proteus vulgaris
  Alcaligenes faecalis

PROCEDURE

Lab One

1. Obtain five of each Purple Broth. Label four of them with the organisms' name, your name, the medium name, and the date. Label one tube of each medium "control."
2. Inoculate four base broths and four of each carbohydrate broth with the test organisms. Do not inoculate the controls.
3. Incubate all tubes aerobically at 35°C for 24 hours.

Lab Two

1. Using the controls for color comparison, examine all tubes and enter your results in the table provided. When indicating the various reactions use the standard symbols as shown with the acid reading first followed by a slash and then the gas reading. K indicates alkalinity and –/– symbolizes no reaction. **Do not try to read these results without control tubes for comparison.**

TABLE 5-3 PURPLE BROTH RESULTS AND INTERPRETATIONS

RESULT	INTERPRETATION	SYMBOL
Yellow broth, bubble in Durham tube	Fermentation with acid and gas end products	A/G
Yellow broth, no bubble in Durham tube	Fermentation with acid end products; no gas produced	A/–
Red broth, no bubble in Durham tube	No fermentation	–/–
Pink broth, no bubble in Durham tube	Degradation of peptone with alkaline end products	K

■ OBSERVATIONS AND INTERPRETATIONS

Enter your results in the table below using the symbols shown in Table 5-3. (See *Photographic Atlas* Figure 6-37.)

OBSERVATIONS AND INTERPRETATIONS

ORGANISM	RESULTS				INTERPRETATION
	PURPLE BASE BROTH	PURPLE GLUCOSE BROTH	PURPLE LACTOSE BROTH	PURPLE SUCROSE BROTH	
Uninoculated Control					

References

DIFCO Laboratories. 1984. Page 660 in *DIFCO Manual*, 10th Ed. DIFCO Laboratories, Detroit, MI.

Forbes, Betty A., Daniel F. Sahm, Alice S. Weissfeld. 2002. Page 269 in *Bailey & Scott's Diagnostic Microbiology*, 11th Ed. Mosby, Inc., St. Louis, MO.

Lányi, B. 1987. Page 44 in *Methods in Microbiology*, Vol. 19, edited by R. R. Colwell and R. Grigorova, Academic Press Inc., New York.

MacFaddin, Jean F. 2000. Page 57 in *Biochemical Tests for Identification of Medical Bacteria*, 3rd Ed. Lippincott Williams & Wilkins, Philadelphia, PA.

Power, David A. and Peggy J. McCuen. 1988. Page 220 in *Manual of BBL® Products and Laboratory Procedures*, 6th Ed. Becton Dickinson Microbiology Systems, Cockeysville, MD.

5-4 METHYL RED AND VOGES-PROSKAUER TESTS

Photographic Atlas Reference
Methyl Red Test Page 67
Voges Proskauer Test Page 80

Materials Needed For This Exercise

Per Student Group

- Three MR-VP broths
- Methyl red
- VP reagents A and B
- Six nonsterile test tubes
- Three nonsterile 1 mL pipettes
- Fresh cultures of:
  - *Escherichia coli*
  - *Enterobacter aerogenes*

Procedure

Lab One

1. Obtain three MR-VP broths. Label two of them with the organisms' names, your name, and the date. Label the third tube "control."
2. Inoculate two broths with the test cultures. Do not inoculate the control.
3. Incubate all tubes at 35°C for 5 days.

Lab Two

1. Transfer 2.0 mL aliquots from each broth (including the control) to two separate tubes. The tubes and pipettes need not be sterile for this. Refer to the procedural diagram in Figure 5-2.
2. For the three pairs of tubes, add the reagents as follows:

Tube #1

a. Add three drops of Methyl Red reagent. Observe for red color formation immediately.
b. Using Table 5-4 as a guide, record your results in the table provided.

Tube #2

a. Add 0.6 mL of VP reagent A. Shake the tubes gently for one minute to oxygenate the medium.
b. Add 0.2 mL of VP reagent B. Shake the tubes again for one minute.
c. Place the tubes in a test tube rack and do not move them for 20 minutes. Observe for red color formation.
d. Using Table 5-4 as a guide, record your results in the table provided. (A common problem with the VP test is differentiating weak positive reactions from negative reactions. VP (–) reactions often produce a copper color; weak VP (+) reactions produce a pink color. The strongest color change should be at the surface.)

TABLE 5-4 RESULTS AND INTERPRETATIONS OF MR-VP TESTS

METHYL RED TEST		
Result	**Interpretation**	**Symbol**
Red	Mixed acid fermentation	+
No color change	No mixed acid fermentation	–
VOGES-PROSKAUER TEST		
Result	**Interpretation**	**Symbol**
Red	Acetoin produced (2,3-butanediol fermentation)	+
No color change	Acetoin is not produced	–

Observations and Interpretations

Refer to Table 5-4 as you record your results and interpretations in the table below.

OBSERVATIONS AND INTERPRETATIONS

Organism	MR Result	VP Result	Interpretation
Uninoculated Control			

■ Figure 5-2 Procedural Diagram for the Methyl Red and Voges-Proskauer Tests

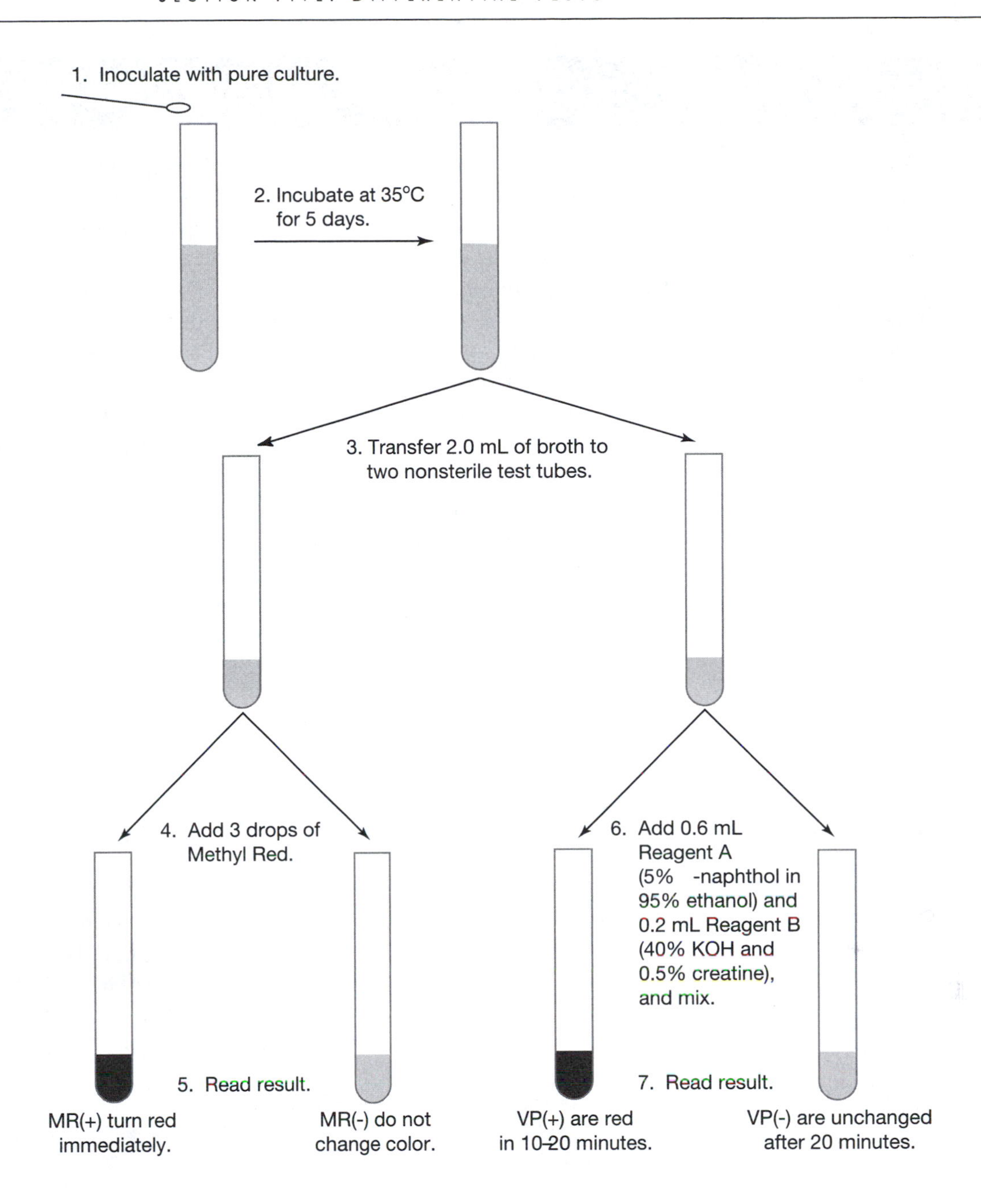

REFERENCES

Delost, Maria Dannessa. 1997. Pages 187–188 in *Introduction to Diagnostic Microbiology.* Mosby, Inc., St. Louis, MO.

DIFCO Laboratories. 1984. Page 543 in *DIFCO Manual*, 10th Ed. DIFCO Laboratories, Detroit, MI.

Forbes, Betty A., Daniel F. Sahm, Alice S. Weissfeld. 2002. Page 275 in *Bailey & Scott's Diagnostic Microbiology*, 11th Ed. Mosby, Inc., St. Louis, MO.

MacFaddin, Jean F. 2000. Pages 321 and 439 in *Biochemical Tests for Identification of Medical Bacteria*, 3rd Ed. Williams & Wilkins, Baltimore, MD.

Power, David A. and Peggy J. McCuen. 1988. Page 202 in *Manual of BBL® Products and Laboratory Procedures*, 6th Ed. Becton Dickinson Microbiology Systems, Cockeysville, MD.

Smibert, Robert M. and Noel R. Krieg. 1994. Pages 622 and 630 in *Methods for General and Molecular Bacteriology*, edited by Philipp Gerhardt, R. G. E. Murray, Willis A. Wood, and Noel R. Krieg, American Society for Microbiology, Washington, DC.

Tests Identifying Microbial Ability to Respire

This unit will examine several techniques designed to differentiate bacteria based on their abilities to respire. As mentioned earlier in this section, respiration is the conversion of glucose to energy in the form of ATP by way of glycolysis, the Krebs cycle and oxidative phosphorylation in the electron transport chain.

Tests that identify an organism as an aerobic or anaerobic respirer are generally designed to detect specific products resulting from reduction of an inorganic final electron acceptor (FEA). Aerobic respirers reduce oxygen to water. Anaerobic respirers reduce other inorganic molecules such as nitrate or sulfate. Nitrate is reduced to nitrogen gas (N_2) or other nitrogenous compounds, while sulfate is reduced to hydrogen sulfide gas (H_2S). Metabolic pathways are described in detail in Appendix A of the *Photographic Atlas*.

The exercises and organisms chosen for this section demonstrate oxidative phosphorylation. The catalase test detects an organism's ability to produce **catalase**, an enzyme that detoxifies hydrogen peroxide produced in the aerobic ETC. The oxidase test identifies the presence of **cytochrome oxidase** in the aerobic electron transport chain. The nitrate reduction test examines bacterial ability to transfer electrons to nitrate at the end of an ETC, and thus respire anaerobically. For more information and biochemical tests that demonstrate anaerobic respiration, refer to TSI and SIM media in "Combination Differential Media," page 105.

5-5 CATALASE TEST

Photographic Atlas Reference
Catalase Test Page 50

MATERIALS NEEDED FOR THIS EXERCISE

Per Student Group

- Three nutrient agar slants
- Hydrogen peroxide (3% solution)
- Transfer pipettes
- Microscope slides
- Fresh cultures of:
  - *Staphylococcus epidermidis*
  - *Enterococcus faecalis*

PROCEDURE

Slide test

1. Transfer a large amount of growth to a microscope slide. (Be sure to perform this test in the proper order. Placing the metal loop into H_2O_2 could catalyze a false positive reaction.)
2. Aseptically place one or two drops of hydrogen peroxide directly onto the bacteria and observe for the formation of bubbles immediately. (When running this test on an unknown, a positive control should be run simultaneously to verify the peroxide's quality.)
3. Record your results in the table provided.

Slant test

1. Add approximately 1 mL of hydrogen peroxide to each of the three slants. Do this *one slant at a time*, observing and recording the results as you go.
2. Record your results in the table provided.

TABLE 5-5 CATALASE TEST RESULTS AND INTERPRETATIONS

RESULT	INTERPRETATION	SYMBOL
Bubbles	Catalase is present	+
No bubbles	Catalase is absent	–

Observations and Interpretations

Using Table 5-5 as a guide, record your results and interpretations in the table below. (See *Photographic Atlas* Figures 6-18 and 6-19.)

Observations and Interpretations			
Organism	**Bubbles?**	**+/-**	**Interpretation**

Observations and Interpretations Slide Test			
Organism	**Bubbles?**	**+/-**	**Interpretation**

Observations and Interpretations Slant Test			
Organism	**Bubbles?**	**+/-**	**Interpretation**

References

Collins, C. H., Patricia M. Lyne, J. M. Grange. 1995. Page 110 in *Collins and Lyne's Microbiological Methods*, 7th Ed., Butterworth-Heinemann, UK.

DIFCO Laboratories. 1984. Page 619 in *DIFCO Manual*, 10th Ed. DIFCO Laboratories, Detroit, MI.

Forbes, Betty A., Daniel F. Sahm, Alice S. Weissfeld. 2002. Pages 151 and 166 in *Bailey & Scott's Diagnostic Microbiology*, 11th Ed. Mosby, Inc., St. Louis, MO.

Lányi, B. 1987. Page 20 in *Methods in Microbiology*, Vol. 19, edited by R. R. Colwell and R. Grigorova, Academic Press Inc., New York.

MacFaddin, Jean F. 2000. Page 78 in *Biochemical Tests for Identification of Medical Bacteria*, 3rd Ed., Williams & Wilkins, Baltimore, MD.

Smibert, Robert M. and Noel R. Krieg. 1994. Page 614 in *Methods for General and Molecular Bacteriology*, edited by Philipp Gerhardt, R. G. E. Murray, Willis A. Wood, and Noel R. Krieg, American Society for Microbiology, Washington, DC.

5-6 OXIDASE TEST

Photographic Atlas Reference
Oxidase Test Page 72

MATERIALS NEEDED FOR THIS EXERCISE

Per Student Group

- Sterile cotton tipped applicator or plastic loop
- Sterile water
- Fresh slant cultures of:
  - *Escherichia coli*
  - *Moraxella catarrhalis*

PROCEDURE (BBL™ *Dry*Slide™ test procedure courtesy of Becton Dickinson and Company.)

1. Following your instructor's guidelines, transfer a small amount of culture to the reagent slide.
2. Observe for color change within 20 seconds.
3. Repeat steps one and two for the second organism.
4. Enter your results in the table provided.
5. When performing this test on an unknown, run a positive control simultaneously to verify the quality of the *Dry*Slide.

REFERENCES

BBL® *Dry*Slide™ package insert.

Collins, C. H., Patricia M. Lyne, J. M. Grange. 1995. Page 116 in *Collins and Lyne's Microbiological Methods*, 7th Ed. Butterworth-Heinemann, UK.

Forbes, Betty A., Daniel F. Sahm, Alice S. Weissfeld. 2002. Pages 151–152 in *Bailey & Scott's Diagnostic Microbiology*, 11th Ed. Mosby, Inc., St. Louis, MO.

Lányi, B. 1987. Page 18 in *Methods in Microbiology*, Vol. 19, edited by R. R. Colwell and R. Grigorova, Academic Press Inc., New York.

MacFaddin, Jean F. 2000. Page 368 in *Biochemical Tests for Identification of Medical Bacteria*, 3rd Ed. Lippincott Williams & Wilkins, Philadelphia, PA.

Smibert, Robert M. and Noel R. Krieg. 1994. Page 625 in *Methods for General and Molecular Bacteriology*, edited by Philipp Gerhardt, R. G. E. Murray, Willis A. Wood, and Noel R. Krieg, American Society for Microbiology, Washington, DC.

TABLE 5-6 OXIDASE TEST RESULTS AND INTERPRETATIONS

Result	Interpretation	Symbol
Dark blue within 20 seconds	Cytochrome oxidase is present	+
No color change to blue or blue after 20 seconds	Cytochrome oxidase is absent	–

■ OBSERVATIONS AND INTERPRETATIONS

Using Table 5-6 as a guide, enter your results in the table below. (See *Photographic Atlas* Figure 6-76)

OBSERVATIONS AND INTERPRETATIONS

Organism	Color Result	+/-	Interpretation
Positive Control			

5-7 Nitrate Reduction Test

Photographic Atlas Reference
Nitrate Reduction Test Page 68
Nitrogen Cycle Page 96

Materials Needed For This Exercise

Per Student Group

- Four nitrate broths
- Nitrate test reagents A and B
- Empty nonsterile test tubes
- Zinc powder
- Fresh cultures of:
  - *Erwinia amylovora*
  - *Escherichia coli*
  - *Pseudomonas aeruginosa*

Procedure

Lab One

1. Obtain four nitrate broths. Label three of them with the organisms' names, your name, and the date. Label the fourth tube "control."
2. Inoculate three broths with the test organisms. Do not inoculate the control.
3. Incubate all tubes at 35°C for 24 to 48 hours.

Lab Two

1. Examine each tube for evidence of gas production. Record your results in the table provided and remove any finished test broths. Refer to Table 5-7 when making your interpretations. (Pay close attention and be methodical when recording your results. Red color can have opposite interpretations depending on where you are in the procedure.)
2. Using Figure 5-3 as a guide proceed to the addition of reagents to *all* tubes not finished (including the control) as follows:
   a. Make a mixture of equal parts reagent A and reagent B. Add approximately 1 mL of the mixture to each tube. Mix well and let the tubes stand undisturbed for 10 minutes. Using the control as a comparison, record your test results in the table provided. Dispose of any test broths that are positive.
   b. Where appropriate add a pinch of zinc dust. Let the tubes stand for 10 minutes. Record your results in the table provided.

References

DIFCO Laboratories. 1984. Page 1023 in *DIFCO Manual*, 10th Ed. DIFCO Laboratories, Detroit, MI.

Forbes, Betty A., Daniel F. Sahm, Alice S. Weissfeld. 2002. Page 277 in *Bailey & Scott's Diagnostic Microbiology*, 11th Ed. Mosby, Inc., St. Louis, MO.

Lányi, B. 1987. Page 21 in *Methods in Microbiology*, Vol. 19, edited by R. R. Colwell and R. Grigorova, Academic Press Inc., New York.

MacFaddin, Jean F. 2000. Page 348 in *Biochemical Tests for Identification of Medical Bacteria*, 3rd Ed. Williams & Wilkins, Philadelphia, PA.

Power, David A. and Peggy J. McCuen. 1988. Page 213 in *Manual of BBL® Products and Laboratory Procedures*, 6th Ed. Becton Dickinson Microbiology Systems, Cockeysville, MD.

Table 5-7 Nitrate Test Results and Interpretations

Result	Interpretation	Symbol
Gas (organism is known to be a nonfermenter)	Denitrification (production of nitrogen gas) $NO_3 \rightarrow NO_2 \rightarrow N_2$	G
Gas (organism's status as fermenter/ nonfermenter is unknown)	Source of gas is unknown; requires addition of reagents	
Red (after addition of reagents A and B)	Nitrate reduction to nitrite $NO_3 \rightarrow NO_2$	+1
No color change (after addition of zinc)	Nitrate reduction to non-gaseous nitrogenous compounds $NO_3 \rightarrow NO_2 \rightarrow$ non gaseous nitrogenous products	+2
Red (after addition of zinc dust)	No nitrate reduction $NO_3 \rightarrow$ no reaction	–

■ Observations and Interpretations

Using Table 5-7 as a guide, enter your results and interpretations in the table below. (See *Photographic Atlas* Figures 6-66 through 6-68.)

OBSERVATIONS AND INTERPRETATIONS				
	Color Results			
Organism	Reagents Added	Zinc Added	Symbol	Interpretation
Uninoculated Control				

■ FIGURE 5-3
Procedural Diagram for the Nitrate Reduction Test

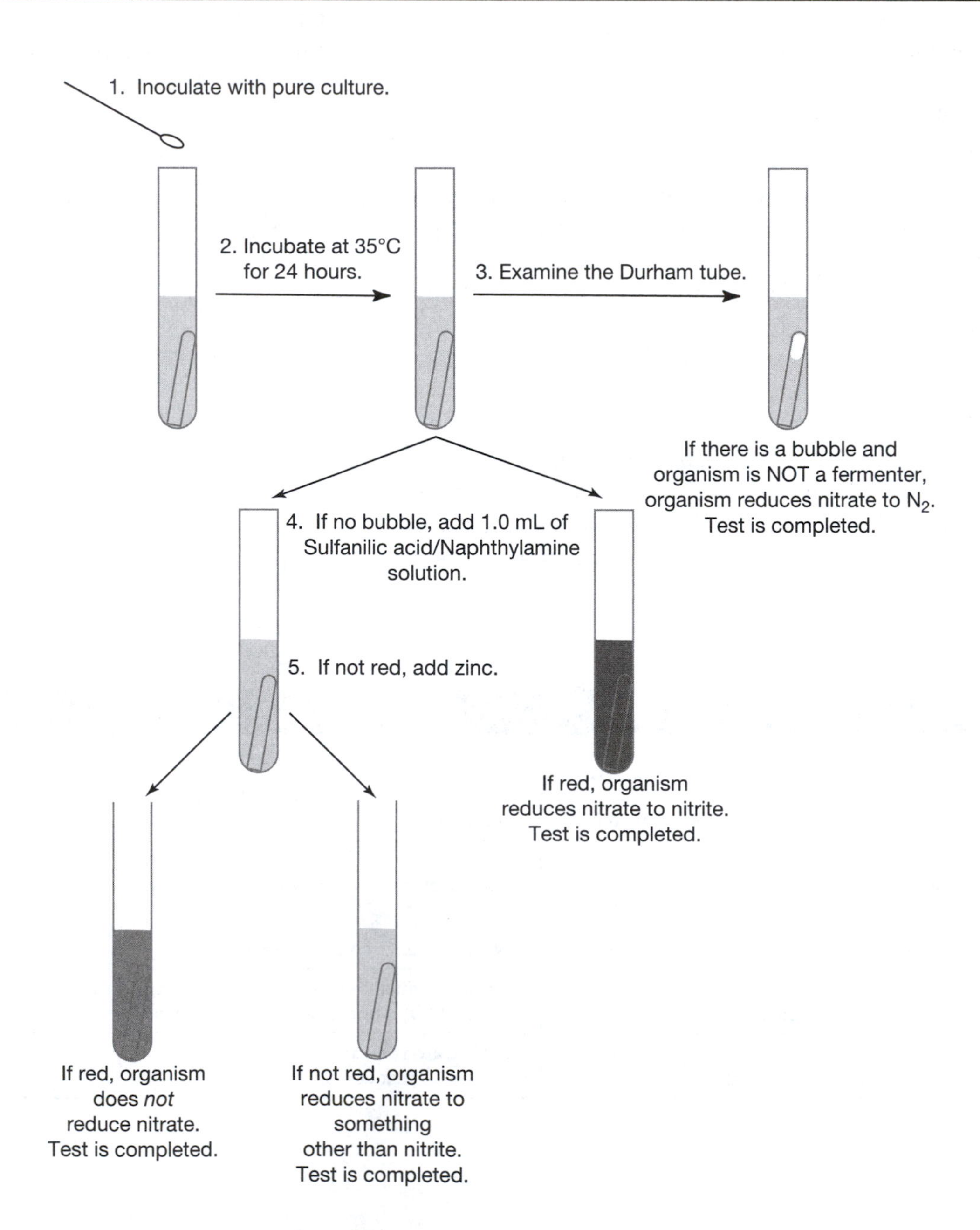

Utilization Media

In this section you will perform utilization tests using two examples of utilization media—Simmons Citrate Medium and Malonate Broth. Utilization media are highly defined formulations designed to differentiate organisms based on their ability to grow when an essential nutrient (*e.g.*, carbon or nitrogen) is available in a limited number of forms. For example, citrate medium contains sodium citrate as the only carbon-containing compound and ammonium ion as the only nitrogen source. Malonate broth contains three sources of carbon but prevents utilization of all but one by competitive inhibition of a specific enzyme.

5-8 CITRATE TEST

Photographic Atlas Reference
Citrate Utilization Test Page 51

MATERIALS NEEDED FOR THIS EXERCISE

Per Student Group

- Three Simmons Citrate slants
- Fresh cultures of:
  Enterobacter aerogenes
  Escherichia coli

PROCEDURE

Lab One

1. Obtain three Simmons Citrate tubes. Label two with the organisms' names, your name, and the date. Label the third tube "control."
2. Using an inoculating needle and *light inoculum*, streak the slants with the test organisms. Do not inoculate the control.
3. Incubate all tubes at 35°C for 48 hours.

Lab Two

1. Observe the tubes for color changes and/or growth.
2. Record your results in the table provided.

REFERENCES

Collins, C. H., Patricia M. Lyne, J. M. Grange. 1995. Page 111 in *Collins and Lyne's Microbiological Methods*, 7th Ed. Butterworth-Heinemann, UK.

DIFCO Laboratories. 1984. Page 864 in *DIFCO Manual*, 10th Ed. DIFCO Laboratories, Detroit, MI.

Forbes, Betty A., Daniel F. Sahm, Alice S. Weissfeld. 2002. Page 266 in *Bailey & Scott's Diagnostic Microbiology*, 11th Ed. Mosby, Inc., St. Louis, MO.

MacFaddin, Jean F. 2000. Page 98 in *Biochemical Tests for Identification of Medical Bacteria*, 3rd Ed. Lippincott Williams & Wilkins, Philadelphia, PA.

Power, David A. and Peggy J. McCuen. 1988. Page 246 in *Manual of BBL® Products and Laboratory Procedures*, 6th Ed. Becton Dickinson Microbiology Systems, Cockeysville, MD.

Smibert, Robert M. and Noel R. Krieg. 1994. Page 614 in *Methods for General and Molecular Bacteriology*, edited by Philipp Gerhardt, R. G. E. Murray, Willis A. Wood, and Noel R. Krieg, American Society for Microbiology, Washington, DC.

TABLE 5-8 CITRATE TEST RESULTS AND INTERPRETATIONS

RESULT	INTERPRETATION	SYMBOL
Blue (even a small amount)	Citrate is utilized	+
No color change; growth	Citrate is utilized	+
No color change; no growth	Citrate is not utilized	–

■ OBSERVATIONS AND INTERPRETATIONS

Using Table 5-8 as a guide, enter your results and interpretations in the table below. (See *Photographic Atlas* Figure 6-21.)

OBSERVATIONS AND INTERPRETATIONS

ORGANISM	COLOR RESULT	+/-	INTERPRETATION
Uninoculated Control			

5-9 MALONATE TEST

Photographic Atlas Reference
Malonate Test Page 66

MATERIALS NEEDED FOR THIS EXERCISE

Per Student Group

- Three Malonate broths
- Fresh cultures of:
  - *Enterobacter aerogenes*
  - *Escherichia coli*

PROCEDURE

Lab One

1. Obtain three Malonate broths. Label two of them with the organisms' names, your name, and the date. Label the third tube "control."
2. Inoculate two broths with the test organisms. Do not inoculate the control.
3. Incubate all tubes at 35°C for 48 hours.

Lab Two

1. Observe the tubes for color changes and record your results in the table provided.

REFERENCES

DIFCO Laboratories. 1984. Page 552 in *DIFCO Manual*, 10th Ed. DIFCO Laboratories, Detroit, MI.

MacFaddin, Jean F. 2000. Page 310 in *Biochemical Tests for Identification of Medical Bacteria*, 3rd Ed. Lippincott Williams & Wilkins, Philadelphia, PA.

Power, David A. and Peggy J. McCuen. 1988. Page 191 in *Manual of BBL® Products and Laboratory Procedures*, 6th Ed. Becton Dickinson Microbiology Systems, Cockeysville, MD.

TABLE 5-9 MALONATE TEST RESULTS AND INTERPRETATIONS

RESULT	INTERPRETATION	SYMBOL
Dark blue	Malonate is utilized	+
No color change or slightly yellow	Malonate is not utilized	–

■ OBSERVATIONS AND INTERPRETATIONS

Using Table 5-9 as a guide, enter your results and interpretations in the table below. (See *Photographic Atlas* Figure 6-58.)

OBSERVATIONS AND INTERPRETATIONS

ORGANISM	COLOR RESULT	+/-	INTERPRETATION
Uninoculated Control			

Decarboxylation and Deamination Tests

Decarboxylation and deamination tests were designed to differentiate between members of *Enterobacteriaceae*. Most members of this group produce one or more enzymes necessary to break down amino acids. Enzymes that catalyze the removal of an amino acid's carboxyl group (COOH) are called **decarboxylases.** Enzymes that catalyze the removal of an amino acid's amine group (NH_2) are called **deaminases.**

Each decarboxylase and deaminase is specific to a particular substrate. Decarboxylases catalyze reactions that produce alkaline products. Thus, identification of an organism's ability to produce a specific decarboxylase is made possible by preparing base medium, adding a known amino acid and including a pH indicator to mark the shift to acidity or alkalinity. Differentiation of an organism in deamination media employs the same principle of substrate exclusivity by including a single known amino acid, but requires the addition of a chemical reagent to produce identifiable colors.

In the next two exercises you will be introduced to the most common of both types of tests. In Exercise 5-10 you will test bacterial ability to decarboxylate arginine, lysine, and ornithine substrates. In Exercise 5-11 you will test bacterial ability to deaminate the amino acid phenylalanine.

5-10 DECARBOXYLATION TEST

Photographic Atlas Reference
Decarboxylation Test Page 54

MATERIALS NEEDED FOR THIS EXERCISE

Per Student Group

- Four Lysine Decarboxylase Broths
- Four Ornithine Decarboxylase Broths
- Four Arginine Decarboxylase Broths
- Four Decarboxylase Base Media
- Sterile mineral oil
- Sterile transfer pipettes
- Recommended organisms:
  Klebsiella pneumoniae
  Enterobacter aerogenes
  Proteus vulgaris

PROCEDURE

Lab One

1. Obtain four of each decarboxylase broth. Label three of each with the organisms' names, your name, and the date. Label the fourth of each broth "control."
2. Lightly inoculate the media with the test organisms. Do not inoculate the control.
3. Overlay all tubes with 3 to 4 mm sterile mineral oil (Figure 5-4).
4. Incubate all tubes aerobically at 35°C for one week.

Lab Two

1. Remove the tubes from the incubator and examine them for color changes.
2. Record your results in the table provided.

TABLE 5-10 DECARBOXYLASE TEST RESULTS AND INTERPRETATIONS

RESULT	INTERPRETATION	SYMBOL
No color change	No decarboxylation	–
Yellow	Fermentation; no decarboxylation	–
Purple (more than the control)	Decarboxylation; organism produces the specific decarboxylase enzyme	+

■ FIGURE 5-4 Adding Mineral Oil to the Tube

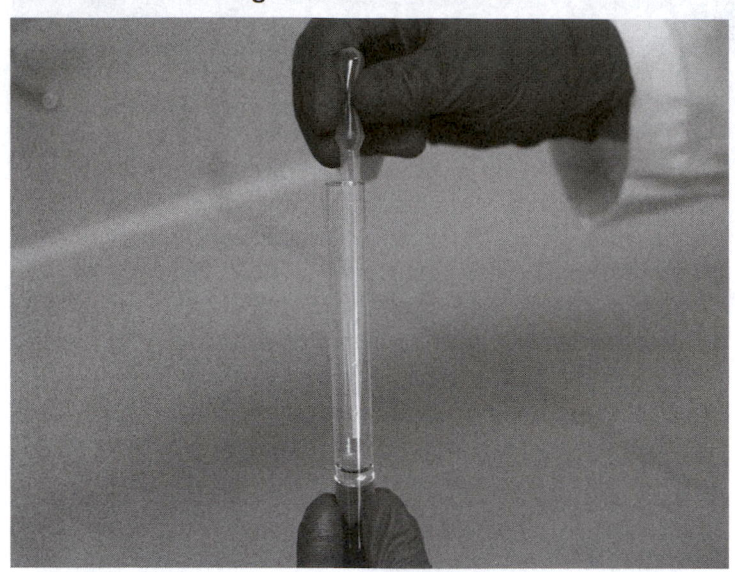

REFERENCES

Collins, C. H., Patricia M. Lyne, J. M. Grange. 1995. Page 111 in *Collins and Lyne's Microbiological Methods*, 7th Ed. Butterworth-Heinemann, UK.

DIFCO Laboratories. 1984. Page 268 in *DIFCO Manual*, 10th Ed. DIFCO Laboratories, Detroit, MI.

Forbes, Betty A., Daniel F. Sahm, Alice S. Weissfeld. 2002. Page 267 in *Bailey & Scott's Diagnostic Microbiology*, 11th Ed. Mosby, Inc., St. Louis, MO.

Lányi, B. 1987. Page 29 in *Methods in Microbiology*, Vol. 19, edited by R. R. Colwell and R. Grigorova, Academic Press Inc., New York.

MacFaddin, Jean F. 2000. Page 120 in *Biochemical Tests for Identification of Medical Bacteria*, 3rd Ed. Lippincott Williams & Wilkins, Philadelphia, PA.

■ OBSERVATIONS AND INTERPRETATIONS

Using Table 5-10 as a guide, record your results and interpretations in the table below. (See *Photographic Atlas* Figures 6-28 through 6-30.)

OBSERVATIONS AND INTERPRETATIONS									
ORGANISM	LYSINE RESULT		ORNITHINE RESULT		ARGININE RESULT		BASE MEDIUM RESULT		INTERPRETATION
	COLOR	+/-	COLOR	+/-	COLOR	+/-	COLOR	+/-	
Uninoculated Control									

5-11 PHENYLALANINE DEAMINASE TEST

Photographic Atlas Reference
Phenylalanine Deaminase Test Page 74

MATERIALS NEEDED FOR THIS EXERCISE

Per Student Group

- Three Phenylalanine Deaminase Agar slants
- 12% ferric chloride solution
- Fresh slant cultures of:
  - *Enterobacter aerogenes*
  - *Proteus vulgaris*

PROCEDURE

Lab One

1. Obtain three Phenylalanine Deaminase slants. Label two slants with the organisms' names, your name, and the date. Label the third slant "control."
2. Streak two slants with heavy inocula of the test organisms. Do not inoculate the control.
3. Incubate all slants aerobically at 35°C for 18 to 24 hours.

Lab Two

1. Add a few drops of 12% ferric chloride solution to each tube and observe for color change. (This color may fade quickly so read and record your results immediately.)
2. Record your results in the table provided.

REFERENCES

DIFCO Laboratories. 1984. Page 664 in *DIFCO Manual*, 10th Ed. DIFCO Laboratories, Detroit, MI.

Forbes, Betty A., Daniel F. Sahm, Alice S. Weissfeld. 2002. Page 280 in *Bailey & Scott's Diagnostic Microbiology*, 11th Ed. Mosby, Inc., St. Louis, MO.

Lányi, B. 1987. Page 28 in *Methods in Microbiology*, Vol. 19, edited by R. R. Colwell and R. Grigorova, Academic Press Inc., New York.

MacFaddin, Jean F. 2000. Page 388 in *Biochemical Tests for Identification of Medical Bacteria*, 3rd Ed. Lippincott Williams & Wilkins, Philadelphia, PA.

Power, David A. and Peggy J. McCuen. 1988. Page 222 in *Manual of BBL® Products and Laboratory Procedures*, 6th Ed. Becton Dickinson Microbiology Systems, Cockeysville, MD.

TABLE 5-11 PHENYLALANINE DEAMINASE TEST RESULTS AND INTERPRETATIONS

RESULT	INTERPRETATION	SYMBOL
Green color	Phenylalanine deaminase is present	+
No color change	Phenylalanine deaminase is absent	–

OBSERVATIONS AND INTERPRETATIONS

Using Table 5-11 as a guide, record your results and interpretations in the table below. (See *Photographic Atlas* Figure 6-81)

OBSERVATIONS AND INTERPRETATIONS

ORGANISM	COLOR RESULT	+/-	INTERPRETATION
Uninoculated Control			

Tests Detecting the Presence of Hydrolytic Enzymes

Reactions that use water to split complex molecules are called **hydrolysis** (or hydrolytic) reactions. The enzymes required for these reactions are called hydrolytic enzymes. When the enzyme catalyzes its reaction inside the cell it is referred to as **intracellular.** Enzymes secreted from the cell to catalyze reactions in the external environment are called **extracellular** enzymes or **exoenzymes.** In this unit you will be performing tests that identify the actions of both types of enzymes.

The hydrolytic enzymes detected by the Urease and Bile Esculin Tests are intracellular and produce identifiable color changes in the medium. Nutrient Gelatin detects gelatinase—an exoenzyme that liquefies the solid medium. DNase Agar, Milk Agar, Tributyrin Agar, and Starch Agar are plated media that identify extracellular enzymes capable of diffusing into the medium and producing a distinguishable halo of clearing around the bacterial growth.

5-12 BILE ESCULIN TEST

Photographic Atlas Reference
Bile Esculin Test Page 47

Materials Needed For This Exercise

Per Student Group

- Three Bile Esculin Agar slants
- Fresh cultures of:
  - *Proteus mirabilis*
  - *Enterococcus faecalis*

Procedure

Lab One

1. Obtain three Bile Esculin Agar slants. Label two with the organisms' names, your name, and the date. Label the third "control."
2. Streak-inoculate two slants with the test organisms. Do not inoculate the control.
3. Incubate all slants at 35°C for up to 48 hours.

Lab Two

1. Observe the slants at 24-hour intervals (if possible) for up to 48 hours and record any that are more than half darkened as positive. Any slants that are less than half darkened are considered negative. Use Table 5-12 as a guide in making your interpretations.
2. Record your results in the table provided.

TABLE 5-12 BILE ESCULIN TEST RESULTS AND INTERPRETATIONS

Result	Interpretation	Symbol
More than half of the medium is darkened within 48 hours	Presumptive identification as a member of group D *Streptococcus* or *Enterococcus*	+
No color change or less than half of the medium is darkened after 48 hours	Presumptive determination as not a member of group D *Streptococcus* or *Enterococcus*	–

Observations and Interpretations

Using Table 5-12 as a guide, record your results and interpretations in the table below. (See *Photographic Atlas* Figure 6-9.)

Observations and Interpretations			
Organism	**Color Result**	**+/-**	**Interpretation**
Uninoculated Control			

References

Delost, Maria Dannessa. 1997. Page 132 in *Introduction to Diagnostic Microbiology.* Mosby, Inc., St. Louis, MO.

DIFCO Laboratories. 1984. Page 129 in *DIFCO Manual*, 10th Ed. DIFCO Laboratories, Detroit, MI.

Forbes, Betty A., Daniel F. Sahm, Alice S. Weissfeld. 2002. Page 264 in *Bailey & Scott's Diagnostic Microbiology*, 11th Ed. Mosby, Inc., St. Louis, MO.

Lányi, B. 1987. Page 56 in *Methods in Microbiology*, Vol. 19, edited by R. R. Colwell and R. Grigorova, Academic Press Inc., New York.

MacFaddin, Jean F. 2000. Page 8 in *Biochemical Tests for Identification of Medical Bacteria*, 3rd Ed. Lippincott Williams & Wilkins, Philadelphia, PA.

Power, David A. and Peggy J. McCuen. 1988. Page 113 in *Manual of BBL® Products and Laboratory Procedures*, 6th Ed. Becton Dickinson Microbiology Systems, Cockeysville, MD.

5-13 STARCH HYDROLYSIS

Photographic Atlas Reference
Starch Hydrolysis Test Page 75

MATERIALS NEEDED FOR THIS EXERCISE

Per Student Group

- One starch agar plate
- Gram iodine (from your Gram stain kit)
- Recommended organisms:
  Bacillus subtilis
  Staphylococcus aureus

PROCEDURE

Lab One

1. Using a marking pen, divide the starch agar plate into three equal sectors. Be sure to mark on the bottom of the plate.
2. Label the plate with the organisms' names, your name, and the date.
3. Spot inoculate two sectors with the test organisms.
4. Invert the plate and incubate it aerobically at 35°C for 48 hours.

Lab Two

1. Remove the plate from the incubator and note the location and appearance of the growth before adding the iodine. (Growth that is thinning at the edge may give the appearance of clearing in the agar after iodine is added to the plate.)
2. Cover the growth and surrounding areas with Gram iodine. Immediately examine the areas surrounding the growth for clearing. (Usually the growth on the agar prevents contact between the starch and iodine so no color reaction takes place at that point. Beginning students sometimes look at this lack of color change and incorrectly judge it as a positive result. Therefore, when examining the agar for clearing, look for a halo *around* the growth, not at the growth itself.)
3. Record your results in the table provided.

REFERENCES

Collins, C. H., Patricia M. Lyne, J. M. Grange. 1995. Page 117 in *Collins and Lyne's Microbiological Methods*, 7th Ed. Butterworth-Heinemann, UK.

DIFCO Laboratories. 1984. Page 879 in *DIFCO Manual*, 10th Ed., DIFCO Laboratories, Detroit, MI.

Lányi, B. 1987. Page 55 in *Methods in Microbiology*, Vol. 19, edited by R. R. Colwell and R. Grigorova, Academic Press Inc., New York, NY.

MacFaddin, Jean F. 2000. Page 412 in *Biochemical Tests for Identification of Medical Bacteria*, 2nd Ed. Lippincott Williams & Wilkins, Philadelphia, PA.

Smibert, Robert M. and Noel R. Krieg. 1994. Page 630 in *Methods for General and Molecular Bacteriology*, edited by Philipp Gerhardt, R. G. E. Murray, Willis A. Wood, and Noel R. Krieg, American Society for Microbiology, Washington, DC.

TABLE 5-13 AMYLASE TEST RESULTS AND INTERPRETATIONS

RESULT	INTERPRETATION	SYMBOL
Clearing around growth	Amylase is present	+
No clearing around growth	No amylase is present	–

OBSERVATIONS AND INTERPRETATIONS

Using Table 5-13 as a guide, enter your results and interpretations in the table below. (See *Photographic Atlas* Figure 6-84.)

OBSERVATIONS AND INTERPRETATIONS

ORGANISM	COLOR RESULT	+/-	INTERPRETATION
Uninoculated Sector			

5-14 UREASE TESTS

Photographic Atlas Reference
Urease Test Page 79

MATERIALS NEEDED FOR THIS EXERCISE

Per Student Group

- Four urease broths
- Four urease agar slants
- Fresh slant cultures of:
  - *Escherichia coli*
  - *Proteus vulgaris*
  - *Klebsiella pneumoniae*

PROCEDURE

Lab One

1. Obtain four of each urease medium. Label three of each with the names of the organisms, your name, and the date. Label the fourth tube of each medium "control."
2. Inoculate three broths with heavy inocula from the test organisms. Do not inoculate the control.
3. Streak-inoculate three slants with the test organisms, covering the entire agar surface with a heavy inoculum. Do not stab the agar butt. Do not inoculate the control.
4. Incubate all tubes aerobically at 35°C for 24 hours. If your labs are scheduled more than 24 hours apart, arrange to have someone place your broths in a refrigerator until you can examine them.

Lab Two

1. Remove all tubes from the incubator and examine them for color changes. Record any rapid positive and slow positive test results in the table provided. Discard all broth tubes and positive agar tubes in an appropriate autoclave container.
2. Return any negative agar tubes to the incubator. Inspect them daily for pink color formation for up to six days. (Negative tubes must not be recorded as negative until they have incubated the full six days.)
3. Enter your agar results daily in the table provided. Use Tables 5-14 and 5-15 as a guide when interpreting your results.

TABLE 5-14 UREASE BROTH TEST RESULTS AND INTERPRETATIONS

RESULT	INTERPRETATION	SYMBOL
Pink	Rapid urea hydrolysis; strong urease production	+
Orange or yellow	No urea hydrolysis; organism does not produce urease or cannot live in broth	-

TABLE 5-15 UREASE AGAR TEST RESULTS AND INTERPRETATIONS

RESULT		INTERPRETATION	SYMBOL
24 HOURS	24 HOURS TO 6 DAYS		
All pink		Rapid urea hydrolysis; strong urease production	+
Partially pink		Slow urea hydrolysis; weak urease production	weak +
Orange or yellow	Partially pink	Slow urea hydrolysis; weak urease production	weak +
Orange or yellow	Orange or yellow	No urea hydrolysis; urease is absent	-

Observations and Interpretations

Using Table 5-14 as a guide, enter your results for the broth test after 24 hours. (See *Photographic Atlas* Figure 6-93.)

OBSERVATIONS AND INTERPRETATIONS UREASE BROTH			
Organism	**Color Result**	**+/-**	**Interpretation**
Uninoculated Control			

Observations and Interpretations

Using Table 5-15 as a guide, enter your results for the agar test each day for 6 days or until pink color appears. (See *Photographic Atlas* Figure 6-92.)

OBSERVATIONS AND INTERPRETATIONS UREASE AGAR								
Organism	**Color**						**+/-**	**Interpretation**
	24 Hrs.	**2 Days**	**3 Days**	**4 Days**	**5 Days**	**6 Days**		
Uninoculated Control								

References

Collins, C. H., Patricia M. Lyne, J. M. Grange. 1995. Page 117 in *Collins and Lyne's Microbiological Methods*, 7th Ed. Butterworth-Heinemann, UK.

Delost, Maria Dannessa. 1997. Page 196 in *Introduction to Diagnostic Microbiology.* Mosby, Inc., St. Louis, MO.

DIFCO Laboratories. 1984. Page 1040 in *DIFCO Manual*, 10th Ed. DIFCO Laboratories, Detroit, MI.

Forbes, Betty A., Daniel F. Sahm, Alice S. Weissfeld. 2002. Page 283 in *Bailey & Scott's Diagnostic Microbiology*, 11th Ed. Mosby, Inc., St. Louis, MO.

Lányi, B. 1987. Page 24 in *Methods in Microbiology*, Vol. 19, edited by R. R. Colwell and R. Grigorova, Academic Press Inc., New York.

MacFaddin, Jean F. 2000. Page 298 in *Biochemical Tests for Identification of Medical Bacteria*, 3rd Ed. Lippincott Williams & Wilkins, Philadelphia, PA.

Power, David A. and Peggy J. McCuen. 1988. Page 280 in *Manual of BBL® Products and Laboratory Procedures*, 6th Ed. Becton Dickinson Microbiology Systems, Cockeysville, MD.

Smibert, Robert M. and Noel R. Krieg. 1994. Page 630 in *Methods for General and Molecular Bacteriology*, edited by Philipp Gerhardt, R. G. E. Murray, Willis A. Wood, and Noel R. Krieg, American Society for Microbiology, Washington, DC.

5-15 CASEASE TEST

Photographic Atlas Reference
Casease Test Page 49

MATERIALS NEEDED FOR THIS EXERCISE

Per Student Group

- One milk agar plate
- Fresh cultures of:
  - *Bacillus subtilis*
  - *Escherichia coli*

PROCEDURE

Lab One

1. Using a marking pen, divide the plate into three equal sectors. Be sure to mark on the bottom of the plate.
2. Label the plate with the organisms' names, your name, and the date.
3. Spot inoculate two sectors with the test organisms and leave the third sector uninoculated as a control.
4. Invert the plate and incubate it aerobically at 35°C for 24 hours.

Lab Two

1. Examine the plates for clearing around the bacterial growth.
2. Record your results in the table provided.

REFERENCES

Chan, E. C. S., Michael J. Pelczar, Jr., and Noel R Krieg. 1986. Page 137 in *Laboratory Exercises In Microbiology*, 5th Ed. McGraw-Hill Book Company.

DIFCO Laboratories. 1984. Page 619 in *DIFCO Manual*, 10th Ed. DIFCO Laboratories, Detroit, MI.

Holt, John G. (Editor). 1994. *Bergey's Manual of Determinative Bacteriology*, 9th Ed. Williams and Wilkins, Baltimore, MD.

Smibert, Robert M. and Noel R. Krieg. 1994. Page 613 in *Methods for General and Molecular Bacteriology*, edited by Philipp Gerhardt, R. G. E. Murray, Willis A. Wood, and Noel R. Krieg, American Society for Microbiology, Washington, DC.

TABLE 5-16 CASEASE TEST RESULTS AND INTERPRETATIONS

RESULT	INTERPRETATION	SYMBOL
Clearing in agar	Casease is present	+
No clearing in agar	Casease is absent	–

OBSERVATIONS AND INTERPRETATIONS

Using Table 5-16 as a guide, record your results in the table below. (See *Photographic Atlas* Figure 6-15.)

OBSERVATIONS AND INTERPRETATIONS

ORGANISM	RESULT	+/-	INTERPRETATION
Uninoculated Sector			

5-16 GELATINASE TEST

Photographic Atlas Reference
Gelatinase Test Page 59

MATERIALS NEEDED FOR THIS EXERCISE

Per Student Group

- Three Nutrient Gelatin stab tubes
- Fresh cultures of:
  - *Bacillus subtilis*
  - *Escherichia coli*

PROCEDURE

Lab One

1. Obtain three Nutrient Gelatin stabs. Label two tubes with the organisms' names, your name, and the date. Label the third tube "control."
2. Stab-inoculate two tubes with heavy inocula of the test organisms. Do not inoculate the control.
3. Incubate all tubes at 25°C for up to one week.

Lab Two

1. Examine the control tube. If the gelatin is solid, the test can be read. If it is liquefied it is likely due to the temperature. All tubes must be refrigerated until the control has solidified.
2. When the control has solidified, examine the inoculated media for gelatin liquefaction.
3. Record your results in the table provided.

REFERENCES

Collins, C. H., Patricia M. Lyne, J. M. Grange. 1995. Page 112 in *Collins and Lyne's Microbiological Methods*, 7th Ed. Butterworth-Heinemann, UK.

DIFCO Laboratories. 1984. Page 35 in *DIFCO Manual*, 10th Ed. DIFCO Laboratories, Detroit, MI.

Forbes, Betty A., Daniel F. Sahm, Alice S. Weissfeld. 2002. Page 271 in *Bailey & Scott's Diagnostic Microbiology*, 11th Ed. Mosby, Inc., St. Louis, MO.

Lányi, B. 1987. Page 44 in *Methods in Microbiology*, Vol. 19, edited by R. R. Colwell and R. Grigorova, Academic Press Inc., New York.

MacFaddin, Jean F. 2000. Page 128 in *Biochemical Tests for Identification of Medical Bacteria*, 3rd Ed. Lippincott Williams & Wilkins, Philadelphia, PA.

Power, David A. and Peggy J. McCuen. 1988. Page 215 in *Manual of BBL® Products and Laboratory Procedures*, 6th Ed. Becton Dickinson Microbiology Systems, Cockeysville, MD.

Smibert, Robert M. and Noel R. Krieg. 1994. Page 617 in *Methods for General and Molecular Bacteriology*, edited by Philipp Gerhardt, R. G. E. Murray, Willis A. Wood, and Noel R. Krieg, American Society for Microbiology, Washington, DC.

TABLE 5-17 GELATINASE TEST RESULTS AND INTERPRETATIONS

RESULT	INTERPRETATION	SYMBOL
Gelatin is liquid (control is solid)	Gelatinase is present	+
Gelatin is solid	No gelatinase is present	–

OBSERVATIONS AND INTERPRETATIONS

Using Table 5-17 as a guide, record your liquefaction results in the table below. (See *Photographic Atlas* Figure 6-40.)

OBSERVATIONS AND INTERPRETATIONS

ORGANISM	RESULT	+/-	INTERPRETATION
Uninoculated Control			

5-17 DNASE TEST

Photographic Atlas Reference
DNase Test Page 55

MATERIALS NEEDED FOR THIS EXERCISE

Per Student Group

- One DNase test agar plate
- Fresh cultures of:
  - *Staphylococcus aureus*
  - *Staphylococcus epidermidis*

PROCEDURE

Lab One

1. Using a marking pen, divide the DNase test agar plate into three equal sectors. Be sure to mark the bottom of the plate.
2. Label the plate with the organisms' names, your name, and the date.
3. Spot inoculate two sectors with the test organisms and leave the third sector as a control.
4. Invert the plate and incubate it aerobically at 35°C for 24 hours. (If your labs are scheduled more than 24 hours apart, arrange to have someone place your plates in a refrigerator until you can examine them.)

Lab Two

1. Examine the plates for clearing around the bacterial growth.
2. Record your results in the table provided.

REFERENCES

Collins, C. H., Patricia M. Lyne, J. M. Grange. 1995. Page 114 in *Collins and Lyne's Microbiological Methods*, 7th Ed. Butterworth-Heinemann, UK.

Delost, Maria Dannessa. 1997. Page 111 in *Introduction to Diagnostic Microbiology.* Mosby, Inc., St. Louis, MO.

DIFCO Laboratories. 1984. Page 263 in *DIFCO Manual*, 10th Ed. DIFCO Laboratories, Detroit, MI.

Forbes, Betty A., Daniel F. Sahm, Alice S. Weissfeld. 2002. Page 268 in *Bailey & Scott's Diagnostic Microbiology*, 11th Ed. Mosby, Inc., St. Louis, MO.

Lányi, B. 1987. Page 33 in *Methods in Microbiology*, Vol. 19, edited by R. R. Colwell and R. Grigorova, Academic Press Inc., New York.

MacFaddin, Jean F. 2000. Page 136 in *Biochemical Tests for Identification of Medical Bacteria*, 3rd Ed. Lippincott Williams & Wilkins, Philadelphia, PA.

Power, David A. and Peggy J. McCuen. 1988. Page 147 in *Manual of BBL® Products and Laboratory Procedures*, 6th Ed. Becton Dickinson Microbiology Systems, Cockeysville, MD.

TABLE 5-18 DNASE TEST RESULTS AND INTERPRETATIONS

RESULT	INTERPRETATION	SYMBOL
Clearing in agar around growth	DNase is present	+
No clearing in agar around growth	DNase is absent	–

■ OBSERVATIONS AND INTERPRETATIONS

Using Table 5-18 as a guide, record your DNase results in the table below. (See *Photographic Atlas* Figure 6-32.)

OBSERVATIONS AND INTERPRETATIONS

ORGANISM	RESULT	+/-	INTERPRETATION
Uninoculated Sector			

5-18 LIPASE TEST

Photographic Atlas Reference
Lipase Test Page 62

Materials Needed For This Exercise

Per Student Group

- One Tributyrin Agar plate
- Fresh cultures of:
  - *Escherichia coli*
  - *Proteus mirabilis*

Procedure

Lab One

1. Using a marking pen, divide the plate into three equal sectors. Be sure to mark on the bottom of the plate. Note: make sure the medium is opaque prior to inoculation. If it is not, obtain a plate that is.
2. Label the plate with the organisms' names, your name, and the date.
3. Spot inoculate two sectors with the test organisms leaving the third sector uninoculated as a control.
4. Invert the plate and incubate it aerobically at 35°C for 24 hours.

Lab Two

1. Examine the plates for clearing around the bacterial growth and record your results in the table provided.

References

Collins, C. H., Patricia M. Lyne, J. M. Grange. 1995. Page 114 in *Collins and Lyne's Microbiological Methods*, 7th Ed. Butterworth-Heinemann, UK.

DIFCO Laboratories. 1984. Page 619 in *DIFCO Manual*, 10th Ed. DIFCO Laboratories, Detroit, MI.

Knapp, Joan S. and Roselyn J. Rice. 1995. Page 335 in *Manual of Clinical Microbiology*, 6th Ed., edited by Patrick R. Murray, Ellen Jo Baron, Michael A. Pfaller, Fred C. Tenover, and Robert H. Yolken, ASM Press, Washington, DC.

MacFaddin, Jean F. 2000. Page 286 in *Biochemical Tests for Identification of Medical Bacteria*, 3rd Ed. Lippincott Williams & Wilkins, Philadelphia, PA.

TABLE 5-19 LIPASE TEST RESULTS AND INTERPRETATIONS

Result	Interpretation	Symbol
Clearing in agar around growth	Lipase is present	+
No clearing in agar around growth	Lipase is absent	–

Observations and Interpretations

Using Table 5-19 as a guide, record your results and interpretations in the table below. (See *Photographic Atlas* Figure 6-48.)

OBSERVATIONS AND INTERPRETATIONS

Organism	Result	+/-	Interpretation
Uninoculated Sector			

Combination Differential Media

All the media included in this unit allow multiple biochemical determinations. Combination media are formulated to include core tests to differentiate members of specific bacterial groups and, when appropriate, can be used to replace a sequence of individual tests. For example, SIM medium tests for sulfur reduction, indole production and motility—tests recommended to differentiate between the genera *Salmonella* and *Shigella*.

As with most biochemical tests, combination test media are typically used in tandem with a *selective* differential medium (Section Four). Selective media promote isolation of desired organisms and provide valuable evidence used to determine the sequence of tests to follow. MacConkey Agar (Exercise 4-6), for example, may be streaked initially to isolate any Gram-negative bacilli followed by the combination medium Kligler's Iron Agar (KIA) or Triple Sugar Iron Agar (TSI). KIA and TSI, which detect fermentation and hydrogen sulfide production, are recommended for differentiation among Gram-negative bacilli.

Other examples of combination media considered in this unit are Lysine Iron Agar (LIA) and Litmus Milk. LIA is used to differentiate enteric bacilli based on their ability to decarboxylate or deaminate lysine, and to respire anaerobically by reducing sulfur. Litmus milk detects up to seven different bacterial biochemical functions, including, lactose fermentation, casein hydrolysis, and deamination.

5-19 SIM MEDIUM

Photographic Atlas Reference
Sulfur Reduction Test Page 76
Indole Production Test Page 60
Motility Test Page 67

MATERIALS NEEDED FOR THIS EXERCISE

Per Student Group

- Three SIM tubes
- Kovac's reagent
- Fresh cultures of:
  - *Escherichia coli*
  - *Proteus mirabilis*

PROCEDURE

Lab One

1. Obtain three SIM tubes. Label two with the organisms' names, your name, and the date. Label third tube "control."
2. Stab-inoculate two tubes with the test organisms. Insert the needle only about two-thirds the depth of the agar. Be careful to remove the needle along the original stab line. Do not inoculate the control.
3. Incubate all tubes aerobically at 35°C for 24 to 48 hours.

Lab Two

1. Examine the tubes for spreading from the stab line *and* formation of black precipitate in the medium. Record any H_2S production and/or motility in the table provided.
2. Add a few drops of Kovac's reagent to each tube. After several minutes observe for the formation of red color in the reagent layer.
3. Record your results in the table provided.

REFERENCES

Delost, Maria Dannessa. 1997. Page 186 in *Introduction to Diagnostic Microbiology.* Mosby, Inc., St. Louis, MO.

MacFaddin, Jean F. 1980. Page 162 in *Biochemical Tests for Identification of Medical Bacteria*, 2nd Ed. Williams & Wilkins, Baltimore, MD.

Power, David A. and Peggy J. McCuen. 1988. Page 246 in *Manual of BBL® Products and Laboratory Procedures*, 6th Ed. Becton Dickinson Microbiology Systems, Cockeysville, MD.

TABLE 5-20 SIM MEDIUM RESULTS AND INTERPRETATIONS SULFUR REDUCTION		
Result	**Interpretation**	**Symbol**
Black in the medium	Sulfur reduction (H_2S production)	+
No black in the medium	Sulfur is not reduced	–
INDOLE PRODUCTION		
Result	**Interpretation**	**Symbol**
Red in the alcohol layer of Kovac's reagent	Tryptophan is broken down into indole and pyruvic acid	+
Reagent color is unchanged	Tryptophan is not broken down into indole and pyruvic acid	–
MOTILITY		
Result	**Interpretation**	**Symbol**
Growth radiating outward from the stab line	Motility	+
No radiating growth	Nonmotile	–

Observations and Interpretations

Using Table 5-20 as a guide, record your results and interpretations in the tables below. (See *Photographic Atlas* Figures 6-44, 6-45, 6-62, 6-63, and 6-88.)

OBSERVATIONS AND INTERPRETATIONS SULFUR REDUCTION			
Organism	**Result**	**+/-**	**Interpretation**
Uninoculated Control			

OBSERVATIONS AND INTERPRETATIONS INDOLE PRODUCTION			
Organism	**Result**	**+/-**	**Interpretation**
Uninoculated Control			

OBSERVATIONS AND INTERPRETATIONS MOTILITY			
ORGANISM	**RESULT**	**+/-**	**INTERPRETATION**
Uninoculated Control			

5-20 TRIPLE SUGAR IRON AGAR

Photographic Atlas Reference
Triple Sugar Iron Agar Page 78

MATERIALS NEEDED FOR THIS EXERCISE

Per Student Group

- Five TSI slants
- Recommended organisms (grown on solid media):
  Pseudomonas aeruginosa
  Escherichia coli
  Morganella morganii
  Proteus vulgaris

PROCEDURE

Lab One

1. Obtain five TSI slants. Label four of the slants with the organisms' names, your name, and the date. Label the fifth slant "control."
2. Inoculate four TSI slants with the test organisms. Using a *heavy* inoculum, stab the agar butt and then streak the slant. Do not inoculate the control.
3. Incubate all slants aerobically at 35°C for 18 to 24 hours.

Lab Two

1. Examine the tubes for characteristic color changes and gas production. Use Table 5-21 as a guide while recording your results in the table provided. The proper format for recording results is: slant reaction/butt reaction, gas production, hydrogen sulfide production. For example, an acid slant and acid butt with gas and black precipitate would be recorded as: A/A, G, H_2S.

REFERENCES

Delost, Maria Dannessa. 1997. Pages 184–185 in *Introduction to Diagnostic Microbiology.* Mosby, Inc., St. Louis, MO.
DIFCO Laboratories. 1984. Page 1019 in *DIFCO Manual*, 10th Ed. DIFCO Laboratories, Detroit, MI.
Forbes, Betty A., Daniel F. Sahm, Alice S. Weissfeld. 2002. Page 282 in *Bailey & Scott's Diagnostic Microbiology*, 11th Ed. Mosby, Inc., St. Louis, MO.
MacFaddin, Jean F. 2000. Page 239 in *Biochemical Tests for Identification of Medical Bacteria*, 3rd Ed. Lippincott Williams & Wilkins, Philadelphia, PA.
Power, David A. and Peggy J. McCuen. 1988. Page 269 in *Manual of BBL® Products and Laboratory Procedures*, 6th Ed. Becton Dickinson Microbiology Systems, Cockeysville, MD.

TABLE 5-21 TSI RESULTS AND INTERPRETATIONS

RESULT	INTERPRETATION	SYMBOL
Yellow slant/yellow butt	Glucose and lactose and/or sucrose fermentation with acid accumulation in slant and butt	A/A
Red slant/yellow butt	Glucose fermentation with acid production; Peptone catabolized aerobically (in the slant) with alkaline products (reversion)	K/A
Red slant/red butt	No fermentation; Peptone catabolized aerobically and anaerobically with alkaline products; Not from *Enterobacteriaceae*	K/K
Red slant /no change in the butt	No fermentation; Peptone catabolized aerobically with alkaline products. Not from *Enterobacteriaceae*	K/NC
No change in slant/no change in butt	Organism is growing slowly or not at all; Not from *Enterobacteriaceae*	NC/NC
Black precipitate in the agar	Sulfur reduction	H_2S
Cracks in or lifting of agar	Gas production	G

Observations and Interpretations

Refer to Table 5-21 when recording and interpreting your results. (See *Photographic Atlas* Figure 6-90.)

OBSERVATIONS AND INTERPRETATIONS			
Organism	**Result**	**Symbol**	**Interpretation**
Uninoculated Control			

5-21 LYSINE IRON AGAR

Photographic Atlas Reference
Lysine Iron Agar Page 65

MATERIALS NEEDED FOR THIS EXERCISE

Per Student Group

- Five LIA slants
- Fresh cultures on solid media:
  - *Citrobacter freundii*
  - *Proteus mirabilis*
  - *Escherichia coli*
  - *Salmonella typhimurium*

PROCEDURE

Lab One

1. Obtain five LIA slants. Label four slants with the organisms' names, your name, and the date. Label the fifth slant "control."
2. Inoculate four slants with the test organisms. Using heavy inocula, stab the agar butt twice and then streak the slant. Do not inoculate the control.
3. Incubate all slants with the caps tightened at 35°C for 18 to 24 hours.

Lab Two

1. Examine the tubes for characteristic color changes.
2. Record your results in the table provided.

REFERENCES

Delost, Maria Dannessa. 1997. Page 194 in *Introduction to Diagnostic Microbiology.* Mosby, Inc., St. Louis, MO.

DIFCO Laboratories. 1984. Page 534 in *DIFCO Manual*, 10th Ed. DIFCO Laboratories, Detroit, MI.

Forbes, Betty A., Daniel F. Sahm, Alice S. Weissfeld. 2002. Page 274 in *Bailey & Scott's Diagnostic Microbiology*, 11th Ed. Mosby, Inc., St. Louis, MO.

Power, David A. and Peggy J. McCuen. 1988. Page 180 in *Manual of BBL® Products and Laboratory Procedures*, 6th Ed. Becton Dickinson Microbiology Systems, Cockeysville, MD.

TABLE 5-22 LIA RESULTS AND INTERPRETATIONS

RESULT	INTERPRETATION	SYMBOL
Purple slant/purple butt	Lysine deaminase negative; Lysine decarboxylase positive	K/K
Purple slant /yellow butt	Lysine deaminase negative; Lysine decarboxylase negative; Glucose fermentation	K/A
Red slant/yellow butt	Lysine deaminase positive; Lysine decarboxylase negative; Glucose fermentation	R/A
Black precipitate	Sulfur reduction	H_2S

OBSERVATIONS AND INTERPRETATIONS

Refer to Table 5-22 when recording and interpreting your results. (See *Photographic Atlas* Figure 6-56.)

OBSERVATIONS AND INTERPRETATIONS

ORGANISM	RESULT	SYMBOL	INTERPRETATION
Uninoculated Control			

5-22 LITMUS MILK MEDIUM

Photographic Atlas Reference
Litmus Milk Medium Page 64

MATERIALS NEEDED FOR THIS EXERCISE

Per Student Group

- Six litmus milk tubes
- Fresh cultures of:
  - *Alcaligenes faecalis*
  - *Pseudomonas aeruginosa*
  - *Klebsiella pneumoniae*
  - *Lactococcus lactis*
  - *Enterococcus faecium*

PROCEDURE

Lab One

1. Obtain six Litmus Milk tubes. Label five tubes with the organisms' names, your name, and the date. Label the sixth tube "control."
2. Inoculate five tubes with the test cultures. Do not inoculate the control.
3. Incubate all tubes aerobically at 35°C for 7 to 14 days.

Lab Two

1. Examine the tubes for color changes, gas production, and clot formation. Be sure to compare all tubes to the control and refer to Table 5-23 when making your interpretations.
2. Record your results in the table provided.

REFERENCES

Forbes, Betty A., Daniel F. Sahm, Alice S. Weissfeld. 2002. Pages 272–273 in *Bailey & Scott's Diagnostic Microbiology*, 11th Ed. Mosby, Inc., St. Louis, MO.

MacFaddin, Jean F. 2000. Page 294 in *Biochemical Tests for Identification of Medical Bacteria*, 3rd Ed. Lippincott Williams & Wilkins, Philadelphia, PA.

Power, David A. and Peggy J. McCuen. 1988. Page 177 in *Manual of BBL® Products and Laboratory Procedures*, 6th Ed. Becton Dickinson Microbiology Systems, Cockeysville, MD.

TABLE 5-23 LITMUS MILK RESULTS AND INTERPRETATIONS

RESULT	INTERPRETATION	SYMBOL
Pink color	Acid reaction	A
Pink and solid (white in the lower portion if the litmus is reduced); clot not movable	Acid clot	AC
Fissures in clot	Gas	G
Clot broken apart	Stormy fermentation	S
White color (lower portion of medium)	Reduction of litmus	R
Semisolid and not pink; clear to gray fluid at top	Curd	C
Clarification of medium; loss of "body"	Digestion of acid clot or curd	D
Blue medium or blue band at top	Alkaline reaction	K
No change	None of the above reactions	NC

(These results may appear together in a variety of combinations.)

Observations and Interpretations

Refer to Table 5-23 when recording and interpreting your results. (See *Photographic Atlas* Figure 6-50.)

OBSERVATIONS AND INTERPRETATIONS			
Organism	**Result**	**Symbol**	**Interpretation**
Uninoculated Control			

Antimicrobial Susceptibility and Resistance

Any microorganism that will grow *in vitro* can be tested for susceptibility to antimicrobial agents. This is an important fact for therapeutic *and* diagnostic reasons. As you learned in the Kirby-Bauer Test (Exercise 2-15), establishing antibiotic susceptibility is an important step in planning a therapeutic course of action. In this unit you will perform two diagnostic tests—Bacitracin Test (for susceptibility) and β-Lactamase Test (for penicillin resistance)—used to differentiate members of *Staphylococcus*, *Streptococcus,* and *Enterococcus*.

5-23 BACITRACIN SUSCEPTIBILITY TEST

Photographic Atlas Reference
Bacitracin Susceptibility Test Page 46

MATERIALS NEEDED FOR THIS EXERCISE

Per Student Group

- One Blood Agar plate (commercial preparation of TSA containing 5% sheep blood)
- Sterile cotton applicators
- 0.04 unit bacitracin disks
- Beaker of alcohol with forceps
- Fresh broth cultures of:
  - *Staphylococcus epidermidis*
  - *Micrococcus roseus*

PROCEDURE

Lab One

1. Obtain one Blood Agar plate. Using a sterile cotton applicator inoculate half of the plate with *S. epidermidis*. (It is important to make the inoculum as light as possible. Do this by wiping and twisting the wet cotton swab on the inside of the culture tube before removing it.) Inoculate the plate by making a single streak nearly halfway across its diameter. Turn the plate 90° and spread the organism evenly so as to produce a bacterial lawn covering half the agar surface.
2. Being careful not to mix the organisms, repeat the process on the other half of the plate using *M. roseus*. Allow the broth to be absorbed by the agar for five minutes before proceeding to step 3.
3. Sterilize the forceps by placing them in the Bunsen burner flame long enough to ignite the alcohol. (Do not hold the forceps in the flame; you are simply burning off the excess alcohol.) Once the alcohol has burned off, use the forceps to place a bacitracin disk in the center of the *S. epidermidis* half of the plate. Gently tap the disk into place to ensure that it makes full contact with the agar surface. Return the forceps to the alcohol.
4. Repeat step 3 placing a bacitracin disk on the *M. roseus* half of the plate. Tap the disk into place and return the forceps to the alcohol.
5. Invert the plate, label it appropriately and incubate it for 24 to 48 hours at room temperature.

Lab Two

1. Remove the plate from the incubator and examine it for clearing around the disks.
2. Record your results in the table provided.

REFERENCES

Baron, Ellen Jo, Lance R. Peterson, and Sydney M. Finegold. 1994. Page 329 in *Bailey & Scott's Diagnostic Microbiology*, 9th Ed. Mosby–Yearbook, Inc., St. Louis, MO.

Delost, Maria Dannessa. 1997. Page 107 in *Introduction to Diagnostic Microbiology.* Mosby, Inc., St. Louis, MO.

DIFCO Laboratories. 1984. Page 292 in *DIFCO Manual*, 10th Ed. DIFCO Laboratories, Detroit, MI.

Forbes, Betty A., Daniel F. Sahm, Alice S. Weissfeld. 2002. Page 290 in *Bailey & Scott's Diagnostic Microbiology*, 11th Ed. Mosby, Inc., St. Louis, MO.

Koneman, Elmer W., *et al.* 1997. Page 551 and 1299 in *Color Atlas and Textbook of Diagnostic Microbiology*, 5th Ed. Lippincott-Raven Publishers, Philadelphia, PA.

MacFaddin, Jean F. 2000. Page 3 in *Biochemical Tests for Identification of Medical Bacteria*, 3rd Ed. Lippincott Williams & Wilkins, Philadelphia, PA.

TABLE 5-24 BACITRACIN TEST RESULTS AND INTERPRETATIONS

RESULT	INTERPRETATION	SYMBOL
Zone of clearing 10 mm or greater	Organism is sensitive to bacitracin	S
Zone of clearing less than 10 mm	Organism is resistant to bacitracin	R

OBSERVATIONS AND INTERPRETATIONS

Refer to Table 5-24 when recording and interpreting your results in the table below. (See *Photographic Atlas* Figure 6-4.)

OBSERVATIONS AND INTERPRETATIONS			
ORGANISM	**RESULT** (ZONE DIAMETER IN MM)	**S/R**	**INTERPRETATION**

5-24 β-LACTAMASE TEST

Photographic Atlas Reference
β-Lactamase Test Page 47

MATERIALS NEEDED FOR THIS EXERCISE

Per Student Group

- Cefinase® discs (available from Becton Dickinson Microbiology Systems, Sparks, MD 21152)
- Fresh slant cultures of:
  - *Staphylococcus epidermidis*
  - *Enterococcus faecalis*

PROCEDURE

1. Place two Cefinase® discs in a sterile Petri dish or on a microscope slide.
2. Inoculate each with a test organism and observe for color changes.

REFERENCES

Delost, Maria Dannessa. 1997. Page 79 in *Introduction to Diagnostic Microbiology.* Mosby, Inc., St. Louis, MO.

Forbes, Betty A., Daniel F. Sahm, Alice S. Weissfeld. 2002. Pages 246–247 in *Bailey & Scott's Diagnostic Microbiology*, 11th Ed. Mosby, Inc., St. Louis, MO.

Koneman, Elmer W., *et al.* 1997. Chapter 15 in *Color Atlas and Textbook of Diagnostic Microbiology*, 5th Ed. Lippincott, Philadelphia, PA.

MacFaddin, Jean F. 2000. Page 254 in *Biochemical Tests for Identification of Medical Bacteria*, 3rd Ed. Lippincott Williams & Wilkins, Philadelphia, PA.

TABLE 5-25 β-LACTAMASE TEST RESULTS AND INTERPRETATIONS

RESULT	INTERPRETATION	SYMBOL
Red or pink	β-lactamase production	+
Yellow or no color change	No β-lactamase production	-

OBSERVATIONS AND INTERPRETATIONS

Refer to Table 5-25 when recording and interpreting your results in the table below. (See *Photographic Atlas* Figure 6-6.)

OBSERVATIONS AND INTERPRETATIONS

ORGANISM	RESULT	+/-	INTERPRETATION

Other Differential Tests

This unit includes tests that do not fit elsewhere but are important tests to consider. Blood Agar is used to cultivate fastidious microorganisms and to detect bacterial (especially *Streptococcus*) hemolytic ability. The coagulase tests are commonly used to presumptively identify pathogenic *Staphylococcus* species. Motility agar is used to detect bacterial motility.

5-25 BLOOD AGAR

Photographic Atlas Reference
Blood Agar Page 48

MATERIALS NEEDED FOR THIS EXERCISE

Per Student Group

- One blood agar plate (Commercially available—TSA containing 5% sheep blood)
- Sterile cotton swabs

PROCEDURE

Lab One

1. Have your lab partner obtain a culture from your throat (Appendix B).
2. Immediately transfer the specimen to a blood agar plate. Use the swab to begin a streak for isolation. Refer to Exercise 1-3 if necessary.
3. Dispose of the swab in a container designated for autoclaving.
4. Complete the streaking with your loop as described in Exercise 1-3.
5. After completing the streak, use your loop to stab the agar in two or three places in the first streak pattern and then in two or three places not previously inoculated.
6. Label the plate with the your name, the specimen source ("throat culture") and the date.
7. Tape the lid down to prevent it from opening accidentally. Invert and incubate the plate aerobically at 25°C for 24 hours.

Lab Two

1. After incubation, do not open your plate until your instructor has seen it and given permission to do so.
2. Observe for color changes and clearing around the isolated growth using transmitted light. This can be done using a colony counter or by holding the plate up to a light. Record your results in the table provided.

REFERENCES

Delost, Maria Dannessa. 1997. Page 103 in *Introduction to Diagnostic Microbiology.* Mosby, Inc., St. Louis, MO.

DIFCO Laboratories. 1984. Page 139 in *DIFCO Manual*, 10th Ed. DIFCO Laboratories, Detroit, MI.

Forbes, Betty A., Daniel F. Sahm, Alice S. Weissfeld. 2002. Page 16 in *Bailey & Scott's Diagnostic Microbiology*, 11th Ed. Mosby, Inc., St. Louis, MO.

Koneman, Elmer W., Stephen D. Allen, William M. Janda, Paul C. Schreckenberger, and Washington C. Winn, Jr. 1997. Chapter 12 in *Color Atlas and Textbook of Diagnostic Microbiology*, 5th Ed. J. B. Lippincott Company, Philadelphia, PA.

Krieg, Noel R. 1994. Page 619 in *Methods for General and Molecular Bacteriology*, edited by Philipp Gerhardt, R. G. E. Murray, Willis A. Wood, and Noel R. Krieg, American Society for Microbiology, Washington, DC.

Power, David A. and Peggy J. McCuen. 1988. Page 115 in *Manual of BBL® Products and Laboratory Procedures*, 6th Ed. Becton Dickinson Microbiology Systems, Cockeysville, MD.

TABLE 5-26 BLOOD AGAR RESULTS AND INTERPRETATIONS

Result	Interpretation	Symbol
Clearing around growth	Organism hemolyzes RBCs completely	β-hemolysis
Greening around growth	Organism hemolyzes RBCs partially	α-hemolysis
No change in the medium	Organism does not hemolyze RBCs	no (γ)-hemolysis

Observations and Interpretations

Refer to Table 5-26 when recording and interpreting your results in the table below. (See *Photographic Atlas* Figure 6-10 through 6-13.)

Observations and Interpretations			
Source of Culture	**Result**	**Symbol**	**Interpretation**

5-26 COAGULASE TESTS

Photographic Atlas Reference
Coagulase Test Page 52

MATERIALS NEEDED FOR THIS EXERCISE

Per Student Group

- Three sterile rabbit plasma tubes (0.5 mL in 12 mm × 75 mm test tubes)
- Sterile 1 mL pipettes
- Sterile saline
- Microscope slides
- Fresh slant cultures of:
  Staphylococcus aureus
  Staphylococcus epidermidis

PROCEDURE

Lab One

Tube Test

1. Obtain three coagulase tubes. Label two tubes with the organisms' names, your name, and the date. Label the third tube "control."
2. Inoculate two tubes with the test organisms. Mix the contents by *gently* rolling the tube between your hands. Do not inoculate the control.
3. Incubate all tubes at 35°C for up to 24 hours, checking for coagulation every 30 minutes for the first 2 to 4 hours.

Slide Test

1. Obtain two microscope slides and divide them into two sides with a marking pen. Label the sides A and B.
2. Place a drop of sterile water on side A and a drop of coagulase plasma on side B of each slide.
3. Transfer a loopful of *S. aureus* to each half of one slide making sure to completely emulsify the bacteria in the solutions. Observe for agglutination within 2 minutes. Clumping after 2 minutes is not a positive result.
4. Repeat step 3 using the other slide and *S. epidermidis*.
5. Record your results in the table provided. Confirm any negative results by comparing with the completed tube test in 24 hours.

Lab Two

1. Remove all tubes from the incubator no later than 24 hours after inoculation. Examine for clotting of the plasma.
2. Record your results in the table provided. Refer to Table 5-27 when making your interpretations.

REFERENCES

Collins, C. H., Patricia M. Lyne, J. M. Grange. 1995. Page 111 in *Collins and Lyne's Microbiological Methods*, 7th Ed. Butterworth-Heinemann, UK.

Delost, Maria Dannessa. 1997. Pages 98–99 in *Introduction to Diagnostic Microbiology.* Mosby, Inc., St. Louis, MO.

DIFCO Laboratories. 1984. Page 232 in *DIFCO Manual*, 10th Ed. DIFCO Laboratories, Detroit, MI.

Forbes, Betty A., Daniel F. Sahm, Alice S. Weissfeld. 2002. Pages 266–267 in *Bailey & Scott's Diagnostic Microbiology*, 11th Ed. Mosby, Inc., St. Louis, MO.

Holt, John G. (Editor). 1994. *Bergey's Manual of Determinative Bacteriology*, 9th Ed. Williams and Wilkins, Baltimore, MD.

Lányi, B. 1987. Page 62 in *Methods in Microbiology*, Vol. 19, edited by R. R. Colwell and R. Grigorova, Academic Press Inc., New York.

MacFaddin, Jean F. 2000. Page 105 in *Biochemical Tests for Identification of Medical Bacteria*, 3rd Ed. Lippincott Williams & Wilkins, Philadelphia, PA.

TABLE 5-27 COAGULASE TEST RESULTS AND INTERPRETATIONS

TUBE TEST

RESULT	INTERPRETATION	SYMBOL
Medium is solid	Plasma has been coagulated	+
Medium is liquid	Plasma has not been coagulated	–

SLIDE TEST

RESULT	INTERPRETATION	SYMBOL
Clumping of cells	Plasma has been coagulated	+
No clumping of cells	Plasma has not been coagulated	–

Observations and Interpretations

Refer to Table 5-27 when recording and interpreting your results in the tables below. (See *Photographic Atlas* Figure 6-22 and 6-23.)

Observations and Interpretations Tube Test Results		
Organism	**Symbol**	**Interpretation**
Uninoculated Control		

Observations and Interpretations Slide Test Results			
Organism	**Side**	**Symbol**	**Interpretation**
	A		
	B		
	A		
	B		

5-27 MOTILITY TEST

Photographic Atlas Reference
Motility Test Page 67

MATERIALS NEEDED FOR THIS EXERCISE

Per Student Group

- Three Motility Test Media stabs
- Fresh cultures of:
  Enterobacter aerogenes
  Klebsiella pneumoniae

PROCEDURE

Lab One

1. Obtain three motility stabs. Label two tubes with the organisms' names, your name, and the date. Label the third tube "control."
2. Stab-inoculate two tubes with the test organisms. (Motility can be obscured by careless stabbing technique. Try to avoid lateral movement when performing this stab.) Do not inoculate the control.
3. Incubate the tubes aerobically at 35°C for 24 to 48 hours.

Lab Two

1. Examine the growth pattern for characteristic spreading from the stab line. Growth will appear red due to an additive in the medium.
2. Record your results in the table provided.

REFERENCES

DIFCO Laboratories. 1984. Page 581 in *DIFCO Manual*, 10th Ed. DIFCO Laboratories, Detroit, MI.

Forbes, Betty A., Daniel F. Sahm, Alice S. Weissfeld. 2002. Page 276 in *Bailey & Scott's Diagnostic Microbiology*, 11th Ed. Mosby, Inc., St. Louis, MO.

MacFaddin, Jean F. 2000. Page 327 in *Biochemical Tests for Identification of Medical Bacteria*, 3rd Ed. Lippincott Williams & Wilkins, Philadelphia, PA.

Power, David A. and Peggy J. McCuen. 1988. Page 201 in *Manual of BBL® Products and Laboratory Procedures*, 6th Ed. Becton Dickinson Microbiology Systems, Cockeysville, MD.

TABLE 5-28 MOTILITY TEST RESULTS AND INTERPRETATIONS

RESULT	INTERPRETATION	SYMBOL
Red diffuse growth radiating outward from the stab line	The organism is motile	+
Red growth only along the stab line	The organism is nonmotile	–

■ OBSERVATIONS AND INTERPRETATIONS

Refer to Table 5-28 when recording and interpreting your results in the table below. (See *Photographic Atlas* Figures 6-62 and 6-63.)

OBSERVATIONS AND INTERPRETATIONS

ORGANISM	RESULT	+/-	INTERPRETATION
Uninoculated Control			

5-28 BACTERIAL UNKNOWNS PROJECT

This exercise is actually a term project to be done when you have completed Sections One through Five. These sections prepare you with the skills needed to successfully do what professional laboratory microbiologists do daily—that is, *isolate* from mixed culture, *grow* in pure culture, and *identify* unknown species of bacteria. Scared? Don't worry. You will not be asked to perform any task not introduced in previous sections. If you have any uncertainty regarding performance of these activities, refer to the appropriate exercises.

You will be given a mixture of two bacteria in broth culture. Your first job is to get the bacteria isolated and growing in pure culture using the techniques of Sections Two and Four. Next, Section Three techniques will be employed as you perform Gram stains on each to determine Gram reaction, and cell morphology and size. Then, results from the differential biochemical tests of Section Five will lead you to identification of your unknowns. And, of course, all work must be done safely and aseptically using the methods covered in Introduction to Safety and Laboratory Guidelines and Section One.

As in Exercise 3-11 (the Morphological Unknown), you will be expected to keep accurate records of all activities, including mistakes and unexpected or equivocal results. You also will be expected to construct a flow chart for each to show the tests you ran and the organisms eliminated by each until you have eliminated all but your unknown. The flow chart is a visual presentation of your thought processes in solving this problem.

While this project may seem intimidating, it is manageable if you don't try to do too much too fast. Take your time and think about what you want to do next based on earlier results. Employ lessons learned in Exercise 3-11 and closely follow the procedures listed below.

This exercise gives you a unique opportunity (and responsibility) to set your own course in a learning project. Take advantage of it; there is great fun and satisfaction to be had if you do. Our experience tells us that students traditionally do one of three things. They:

- treat the project as a fun puzzle to solve,
- become completely stressed out and hate every minute of it,
- get lost early on and, for whatever reason, don't ask for help until it is too late to receive meaningful assistance.

Many of you will be nervous, perhaps even confused at first; your instructor understands this. This is normal and can actually help you to focus if you use it in a positive way. For your success and overall satisfaction with this project, *ask your instructor for help if you need it*. Happy hunting!

MATERIALS

- Recommended organisms (to be mixed in pairs in a broth by the lab technician immediately prior to use—generally, a few drops of the Gram-negative added to an overnight culture of the Gram-positive provides an appropriate ratio for isolation and growth):
  - Gram-positives
    - *Bacillus cereus*
    - *Bacillus coagulans*
    - *Corynebacterium xerosis*
    - *Enterococcus faecalis*
    - *Lactobacillus plantarum*
    - *Lactococcus lactis*
    - *Micrococcus luteus*
    - *Micrococcus roseus*
    - *Mycobacterium smegmatis*
    - *Staphylococcus aureus*
    - *Staphylococcus epidermidis*
  - Gram-negatives
    - *Aeromonas hydrophila*
    - *Alcaligenes faecalis*
    - *Chromobacterium violaceum*
    - *Citrobacter amalonaticus*
    - *Enterobacter aerogenes*
    - *Erwinia amylovora*
    - *Escherichia coli*
    - *Hafnia alvei*
    - *Moraxella catarrhalis*
    - *Morganella morganii*
    - *Neisseria sicca*
    - *Proteus mirabilis*
    - *Pseudomonas aeruginosa*
- Appropriate stains and biochemical media

PROCEDURE

1. Preliminary duties
   a. Your instructor will tell you which organisms will actually be used based on your lab's inventory. You will also be advised as to the optimum temperature for the particular strain your lab has of each organism.
   b. Your instructor will tell you which media and stains will be available for testing.
   c. Working as a class, you will determine the results of each organism for each test available. This information will provide a database of results that you can use to compare against your unknown's results. We recommend running and using these "class controls" in lieu of referring to results in *Bergey's Manual of Systematic Bacteriology* or some other standard reference for the following reasons
      - as many as 10% of the strains of a species listed as positive or negative on a test give the opposite result

- many species have even higher strain variability
- not all test results are listed for all organisms

d. Class control tests should be run for the standard times, unless the timing of class sessions makes this impractical. It is imperative that incubation times for tests on unknowns be exactly the same as was used for the controls.

e. Incubate your class controls at the optimum temperature for each species as given to you by your instructor.

f. Class control results will be collected, tabulated, and distributed to each student. Your instructor will provide details on this process.

2. Isolation of the Unknown

a. You will be given a broth containing a *fresh* mixture of two unknown bacteria selected from the list of possible organisms. Enter the number of your unknown on your Data Sheets.

b. Mix the broth, then streak for isolation on two agar plates. Your instructor will tell you what media are available for use. You may be supplied with an undefined medium, such as trypticase soy agar, or a selective medium, such as phenylethyl alcohol agar or desoxycholate agar. Enter all relevant information concerning your isolation procedures on the Data Sheets, including date, medium, source of bacteria, and incubation temperature.

c. Incubate one agar plate at 25°C and the other at 37°C for at least 24 hours.

d. After incubation, check for isolated colonies that have different morphologies. If you have isolation of both, go on to step "3a." If you do *not* have isolation of *either*, continue with step "2e." If you have isolation of only one, go to step "3a" for the isolated species and step "2e" for the one not isolated. Be sure to enter relevant information on the Data Sheets at the end of this section.

e. If you do not have isolation, follow the advice that best matches your situation. (Be sure to record everything you do in the isolation. Include the date, source of inoculum, medium to which it is being transferred, and incubation temperature.)

- Look for growth with different colony morphologies, even if they're not separate. If you see different growth, use a portion of each and streak more plates. Then incubate them for either a shorter time or at a suboptimal temperature, since they grew *too* well the first time.
- If you see only one type of growth, ask for a selective medium that favors growth of the one you're missing. (This will require a Gram stain of the one you do have.) Streak the mixture and incubate again. Also, reincubate your original plate—some species are slow growers, so their absence may be due to a slow growth rate.

After incubation, observe the plate(s) and repeat or go to step "3a", whichever is appropriate. You may also consult with your instructor for guidance in particularly difficult situations.

3. Growing the Unknown in Pure Culture

a. Once you have isolation of an unknown, transfer a portion of its colony to an agar slant to produce a pure culture. Use the rest of the colony for a Gram stain. (If you don't have enough colony left for a Gram stain, do the Gram stain on your pure culture after incubation.)

b. After Gram staining, label your pure culture accordingly.

c. Enter the following information on your Data Sheet: optimum growth temperature, colony morphology, cell morphology and arrangement, and cell size. Also, complete the description of your isolation procedure by noting the source of the isolate, the medium to which you are transferring it, and the incubation temperature.

4. Identification of the Unknown

a. Follow this procedure for each unknown. Be sure to enter inoculation and reading dates for each test on the Data Sheet. Also include the test result and any comments about the test that explain any deviation from standard procedure (*i.e.,* reading tests before or after the accepted incubation time, running a test and not using it in the flow chart, *etc.*).

b. Construct a flow chart that divides all the organisms up first by Gram reaction, then by cellular morphology. Use the flow chart in Figure 5-5 as a style guide even though the actual organisms you use will be different.

c. There are two options for proceeding at this point. Your instructor will let you know which to use.

1) Find the group of organisms on the flow chart that matches the results of your unknown. Then, look at your class controls results for *just those organisms* and choose a test that will divide these organisms into at least two groups. (Also consider your ability to return and read the test after the appropriate incubation time. That is, don't inoculate a 48-hour test on Thursday if you can't get into the lab on Saturday!) Continue the flow chart from the branch with the remaining candidates for your unknown.

2) Rather than allowing you to choose just any test that works, you may be required to follow some form of standard approach to identification. Your instructor will provide you with relevant information based on the organism inventory you are working with.

d. Inoculate the test medium you have chosen and incubate it for the appropriate time. The optimum temperature for growth on your streak plates will be used as the incubation temperature. *It is important to run your tests at that optimum temperature because this is the temperature at which the class controls were run.*

e. While the test is incubating, you should begin planning what your next test will be. A final decision about the next test cannot be made until you have results from the first, but you can decide which test to run if the result is positive, and which one to run if it is negative. This applies to all stages of the flow chart: since you won't know if a subsequent test is relevant or not until you have results from the current one, *you should not inoculate a medium until you have those results.* It's really easy: run one test at a time for each unknown. (Note: In a clinical situation, where rapid identification of a pathogen may save a patient's life, tests are routinely run concurrently. However, remember that correct identification is only one objective of this project. More important is for you to demonstrate an understanding of the logic behind the process and execute it in the most efficient manner.)

f. Repeat the process of inoculating a medium, getting the results, and then choosing a subsequent test until you eliminate all but one organism. This *should* be your unknown. Then continue with step "4g."

g. When you have eliminated all but one organism, you will run one more test—the confirmatory test. This must be one that has not been run previously on your organism. It's also nice (but not necessary) if the test you choose gives a positive result (since, in general, we have more confidence in positive results than in negative results, because false positives are usually harder to get than false negatives.) The confirmatory test provides you with the unique opportunity to predict the result before you run the test. If it matches, you are more certain that you have correctly identified your unknown. Continue with step "4j." If it doesn't match, continue with step "4h."

h. If your confirmatory test doesn't match the result you expected for your unknown, check with your instructor to see where your organism was eliminated incorrectly. In most cases, it will be difficult to determine at this point what was responsible—you, the class controls, or the organism itself. Misidentification may be due to one or any combination of factors:
   - the test procedure could have been done incorrectly by you or the person responsible for running class controls on your organism
   - the test may have been interpreted incorrectly by you or the person responsible for running class controls on your organism
   - the inoculum was too small in your test or the class controls to give a positive result in the limited incubation time
   - the wrong organism was inoculated at the time of the test or the class controls (many look alike in a tube, and once the label goes on, as far as the microbiologist is concerned, that culture becomes the labeled organism regardless of what's really in there!)
   - for whatever reason, the organism just didn't react "correctly" during your test or during the class controls.

i. Based on your instructor's advice, you will either
   - rerun the test where you incorrectly eliminated your unknown (and perhaps rerun the test on the remaining organisms to check the class control results), or
   - rerun the test where you incorrectly eliminated your unknown without running the controls again, or
   - eliminate the problematic test from your flow chart, but use the other tests you've already done. If these don't allow you to identify your unknown, more tests will need to be run. Continue with step "4f."

   NOTE: There will undoubtedly be some "mistakes" uncovered in the class controls as they get repeated. These need to be reported to the class so the correct information can be incorporated into their flow charts.

j. When you have correctly identified your unknown, complete the Data Sheet and turn it in. Your instructor will advise you as to the point value of each section, the grading scale, and any other items that are required.

■ FIGURE 5-5 Sample Flow Chart

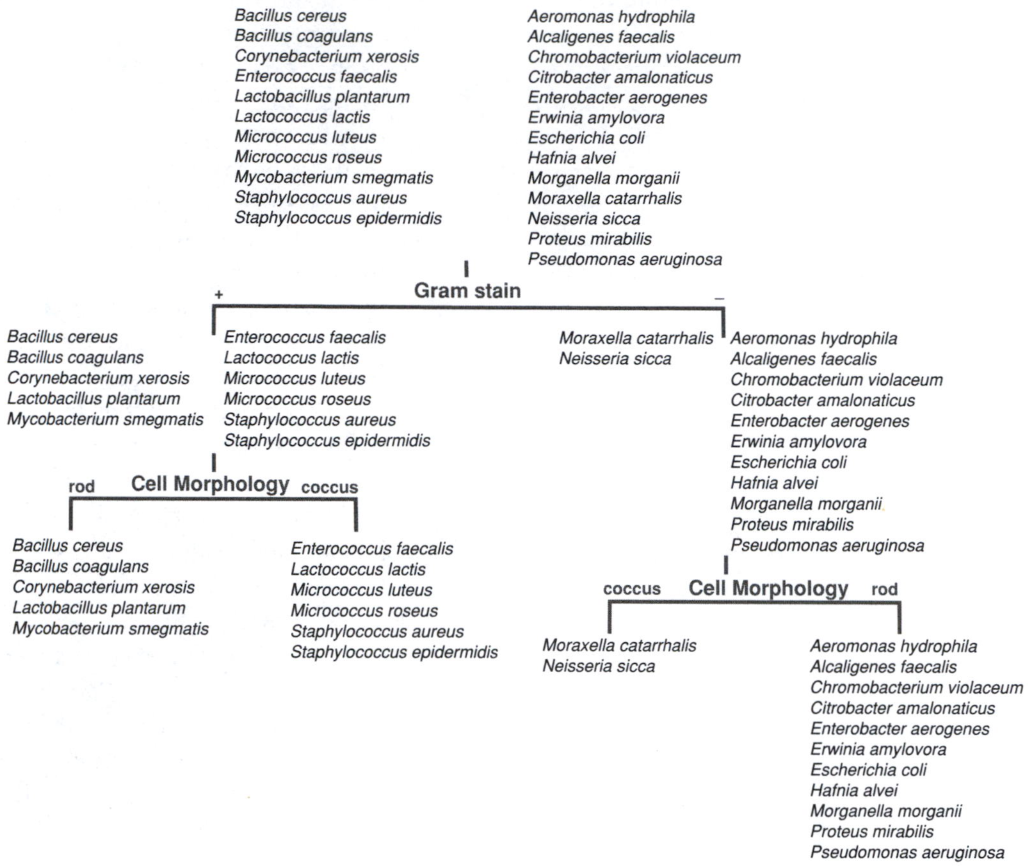

Bacillus cereus
Bacillus coagulans
Corynebacterium xerosis
Enterococcus faecalis
Lactobacillus plantarum
Lactococcus lactis
Micrococcus luteus
Micrococcus roseus
Mycobacterium smegmatis
Staphylococcus aureus
Staphylococcus epidermidis

Aeromonas hydrophila
Alcaligenes faecalis
Chromobacterium violaceum
Citrobacter amalonaticus
Enterobacter aerogenes
Erwinia amylovora
Escherichia coli
Hafnia alvei
Morganella morganii
Moraxella catarrhalis
Neisseria sicca
Proteus mirabilis
Pseudomonas aeruginosa

Gram stain

+

- *Bacillus cereus*, *Bacillus coagulans*, *Corynebacterium xerosis*, *Lactobacillus plantarum*, *Mycobacterium smegmatis*, *Enterococcus faecalis*, *Lactococcus lactis*, *Micrococcus luteus*, *Micrococcus roseus*, *Staphylococcus aureus*, *Staphylococcus epidermidis*
  - **Cell Morphology**
    - rod: *Bacillus cereus*, *Bacillus coagulans*, *Corynebacterium xerosis*, *Lactobacillus plantarum*, *Mycobacterium smegmatis*
    - coccus: *Enterococcus faecalis*, *Lactococcus lactis*, *Micrococcus luteus*, *Micrococcus roseus*, *Staphylococcus aureus*, *Staphylococcus epidermidis*

–

- *Moraxella catarrhalis*, *Neisseria sicca*, *Aeromonas hydrophila*, *Alcaligenes faecalis*, *Chromobacterium violaceum*, *Citrobacter amalonaticus*, *Enterobacter aerogenes*, *Erwinia amylovora*, *Escherichia coli*, *Hafnia alvei*, *Morganella morganii*, *Proteus mirabilis*, *Pseudomonas aeruginosa*
  - **Cell Morphology**
    - coccus: *Moraxella catarrhalis*, *Neisseria sicca*
    - rod: *Aeromonas hydrophila*, *Alcaligenes faecalis*, *Chromobacterium violaceum*, *Citrobacter amalonaticus*, *Enterobacter aerogenes*, *Erwinia amylovora*, *Escherichia coli*, *Hafnia alvei*, *Morganella morganii*, *Proteus mirabilis*, *Pseudomonas aeruginosa*

BACTERIAL UNKNOWNS PROJECT

GRAM POSITIVE UNKNOWN DATA SHEET

Name ______________________________

Date ______________________ Lab Section ______________________

UNKNOWN NUMBER _______

ISOLATION PROCEDURE (Please record all activities associated with isolation of your organisms—from mixed culture to pure culture. Always include the date, source of inoculum, destination, incubation temperature, and any other relevant information. Also make note of transfers made to keep your pure culture fresh. This log must be kept current.)

____

____

____

____

____

____

____

____

____

____

____

____

____

____

____

____

____

____

PRELIMINARY OBSERVATIONS

Colony Morphology ______________________________

____

Gram Stain ______________ Cellular Morphology and Arrangement ______________

Cell Dimensions ______________________ Optimum Temperature ______________

DIFFERENTIAL TESTS (Please begin recording with your first successful Gram stain. Include all information through the confirmatory test. This log must be kept current.)

Test #1:__________ Date Begun:_______ Date Read: _______ Result: __________
Comments: ______________________________

Test #2:__________ Date Begun:_______ Date Read: _______ Result: __________
Comments: ______________________________

Test #3:__________ Date Begun:_______ Date Read: _______ Result: __________
Comments: ______________________________

Test #4:__________ Date Begun:_______ Date Read: _______ Result: __________
Comments: ______________________________

Test #5:__________ Date Begun:_______ Date Read: _______ Result: __________
Comments: ______________________________

Test #6:__________ Date Begun:_______ Date Read: _______ Result: __________
Comments: ______________________________

Test #7:__________ Date Begun:_______ Date Read: _______ Result: __________
Comments: ______________________________

Test #8:__________ Date Begun:_______ Date Read: _______ Result: __________
Comments: ______________________________

Test #9:__________ Date Begun:_______ Date Read: _______ Result: __________
Comments: ______________________________

Test #10: __________ Date Begun:_______ Date Read: _______ Result: __________
Comments: ______________________________

Test #11: __________ Date Begun:_______ Date Read: _______ Result: __________
Comments: ______________________________

Test #12: __________ Date Begun:_______ Date Read: _______ Result: __________
Comments: ______________________________

Test #13: __________ Date Begun:_______ Date Read: _______ Result: __________
Comments: ______________________________

Test #14: __________ Date Begun:_______ Date Read: _______ Result: __________
Comments: ______________________________

MY UNKNOWN IS: ______________________________

____

____

BACTERIAL UNKNOWNS PROJECT

GRAM NEGATIVE UNKNOWN DATA SHEET

Name ______________________________

Date ______________________ Lab Section ______________________

UNKNOWN NUMBER ________

ISOLATION PROCEDURE (Please record all activities associated with isolation of your organisms—from mixed culture to pure culture. Always include the date, source of inoculum, destination, incubation temperature, and any other relevant information. Also make note of transfers made to keep your pure culture fresh. This log must be kept current.)

PRELIMINARY OBSERVATIONS

Colony Morphology ______________________________

Gram Stain ______________ Cellular Morphology and Arrangement ______________

Cell Dimensions ______________________ Optimum Temperature ______________

DIFFERENTIAL TESTS (Please begin recording with your first successful Gram stain. Include all information through the confirmatory test. This log must be kept current.)

Test #1:__________ Date Begun:__________ Date Read:__________ Result:__________
Comments:__________

Test #2:__________ Date Begun:__________ Date Read:__________ Result:__________
Comments:__________

Test #3:__________ Date Begun:__________ Date Read:__________ Result:__________
Comments:__________

Test #4:__________ Date Begun:__________ Date Read:__________ Result:__________
Comments:__________

Test #5:__________ Date Begun:__________ Date Read:__________ Result:__________
Comments:__________

Test #6:__________ Date Begun:__________ Date Read:__________ Result:__________
Comments:__________

Test #7:__________ Date Begun:__________ Date Read:__________ Result:__________
Comments:__________

Test #8:__________ Date Begun:__________ Date Read:__________ Result:__________
Comments:__________

Test #9:__________ Date Begun:__________ Date Read:__________ Result:__________
Comments:__________

Test #10:__________ Date Begun:__________ Date Read:__________ Result:__________
Comments:__________

Test #11:__________ Date Begun:__________ Date Read:__________ Result:__________
Comments:__________

Test #12:__________ Date Begun:__________ Date Read:__________ Result:__________
Comments:__________

Test #13:__________ Date Begun:__________ Date Read:__________ Result:__________
Comments:__________

Test #14:__________ Date Begun:__________ Date Read:__________ Result:__________
Comments:__________

MY UNKNOWN IS: __________

____

____

SECTION SIX

6 Quantitative Techniques

Microbiological quantitative techniques, simply defined, are the methods used to estimate the number of microorganisms or viruses per unit volume (*density*) in a given specimen. Of the five techniques examined in this section, only Direct Count (Exercise 6-3) involves the actual counting of cells. The other four employ a variety of techniques to *indirectly* measure density. The turbidimetric technique used in Exercise 6-5 Closed System Growth may be used for quantitative purposes. We offer it simply as an exercise in examining the phases of bacterial growth under a variety of temperature conditions.

As you proceed through the exercises in this section you will gain understanding and proficiency with **serial dilutions.** In microbiology, as well as other areas of science, a working knowledge of serial dilutions and dilution factors is essential. One need only try to imagine the impossible task of counting bacteria in a broth to understand why systematically diluting it is beneficial.

As you undoubtedly predicted, serial dilutions involve math. Although the math may seem difficult at first it can be easily understood with a little study and a lot of *practice*. This cannot be overemphasized! Ask your instructor to give you practice problems to help you become proficient with the formulas. Prepare yourself by reading the exercises well before your lab session and, if possible, discuss them with members of your lab group. Ask your instructor for help if there are aspects of an exercise you don't understand, preferably *before class*. You will have a much more pleasant lab experience if you do.

6-1 STANDARD PLATE COUNT (VIABLE COUNT)

Photographic Atlas Reference
Standard Plate Count Page 83

MATERIALS NEEDED FOR THIS EXERCISE

Per Student Group

- Micropipettes (10–100 µL and 100–1000 µL) with sterile tips
- Sterile microtubes
- Flask of sterile water
- Eight nutrient agar plates
- Beaker containing ethanol and a bent glass rod
- Hand tally counter
- Colony counter
- 24 hour broth culture of *Escherichia coli*

PROCEDURE (Refer to the procedural diagram in Figure 6-1 and Exercise 1-4 as needed.)

Lab One

1. Obtain eight plates, organize them into 4 pairs and label them A_1, A_2, B_1, B_2, *etc.*
2. Obtain 5 microtubes and label them 1–5. These are your dilution tubes; make sure they remain covered until needed.
3. Aseptically add 990 µL sterile water to dilution tubes 1 and 2. Cover when finished. Aseptically add 900 µL sterile water to dilution tubes 3, 4, and 5. Cover when finished.
4. Mix the broth culture and aseptically transfer 10 µL to dilution tube 1; mix well. This is dilution factor 10^{-2} (DF 10^{-2}).
5. Aseptically transfer 10 µL from dilution tube 1 to dilution tube 2; mix well. This is DF 10^{-4}.
6. Aseptically transfer 100 µL from dilution tube 2 to dilution tube 3; mix well. This is DF 10^{-5}.
7. Aseptically transfer 100 µL from dilution tube 3 to dilution tube 4; mix well. This is DF 10^{-6}.
8. Aseptically transfer 100 µL from dilution tube 4 to dilution tube 5; mix well. This is DF 10^{-7}.
9. Aseptically transfer 100 µL from dilution tube 2 to plate A_1. Using the spread plate technique disperse the diluent evenly over the entire surface of the agar. Repeat the procedure with plate A_2 and label both plates "FDF 10^{-5}". FDF (final dilution factor)* is explained below.

* **Final Dilution Factors (FDF).** Cell density is calculated in CFU/mL based on the number of colonies produced on an agar plate. Standard count plates, however, are typically inoculated with only 0.1 mL (100 µL) of the diluted sample. If you do not account for this discrepancy the calculated original density will be off by a factor of ten. Therefore, since the plating of 0.1 mL dilution will yield 1/10th as many colonies as 1.0 mL would, the act of plating itself is treated as another tenfold dilution. (In rare instances, where a full milliliter (1000 µL) is plated, the dilution factor of the plate remains the same as the source tube.) So if the dilution factor of the source tube is 10^{-7}, the FDF written on the plate will be 10^{-8}. In this way, the original density in CFU per milliliter can be simply determined by dividing the number of colonies by the FDF.

FIGURE 6-1 Serial Dilution Procedural Diagram
This is an illustration of the dilution scheme outlined in the Procedure. Use it as a guide while following the outlined procedure. As you perform the series of dilutions you will be assigning "dilution factors" to the tubes and their contents. These dilution factors indicate the proportion of original culture broth present in the dilution tube. Note that the final dilution factor of the plates is ten times greater than that in the source tube. This is because cell density is calculated in CFU/mL (or CFU/1000 µL), whereas plates typically receive only 0.1 mL (100 µL) of liquid. Refer to *Photographic Atlas* page 83 for a full explanation.

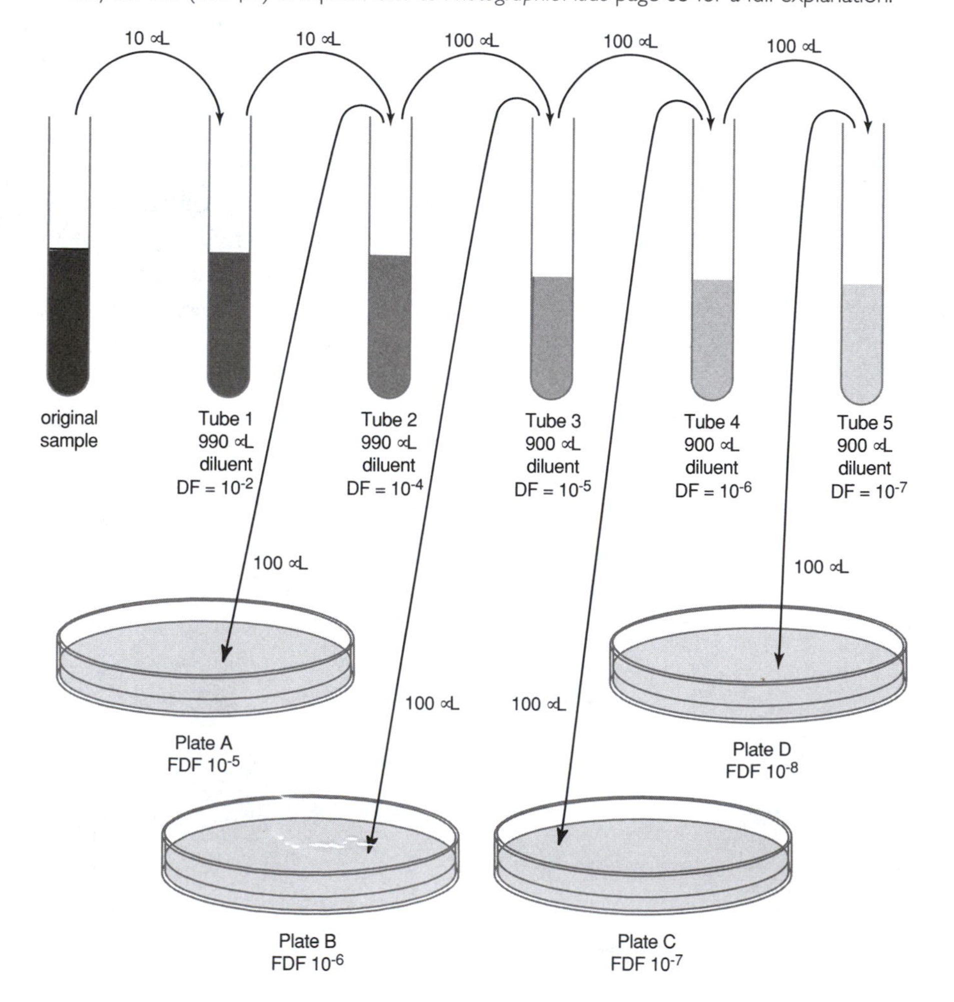

10. Following the same procedure, transfer 100 µL volumes from dilution tubes 3, 4, and 5 to plates B, C, and D respectively. Label the plates with their appropriate FDF.
11. Invert the plates and incubate at 35°C for 24 to 48 hours.

Lab Two

1. After incubation examine the plates and determine the countable pair—plates with 30 to 300 colonies. Only one pair of plates *should* be countable.
2. Count the colonies on both plates and calculate the average (Figure 6-2). Record it in the table provided. (Note: due to error, you may have more than one pair that is countable. Count *all* plates that have between 30 and 300 colonies for the practice and try to identify which plate(s) you have the most confidence in. If *no* plates are in the 30–300 colony range, count the pair that is closest just for the practice and for purposes of the calculations.
3. Using the formula provided in Data and Calculations calculate the density of the original sample and record it in the space provided.

FIGURE 6-2 Counting Bacterial Colonies
Place the plate upside down on the colony counter. Turn on the light and adjust the magnifying glass until all the colonies are visible. Using the grid in the background as a guide, count the colonies one section at a time. Mark each colony with a felt-tip marker as you record with a hand tally counter.

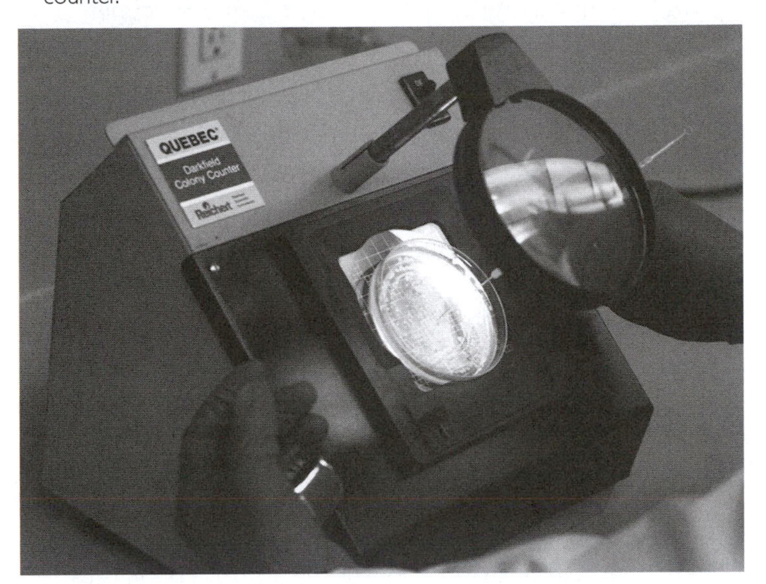

DATA AND CALCULATIONS

1. Enter the number of colonies counted on each countable plate. Only one pair of plates should be countable, but for practice, record all countable plates anyway. For all plates containing greater than 300 colonies, enter TMTC (too many to count). For plates containing fewer than 30, enter TFTC (too few to count).
2. Take the average number of colonies from the two (or more) countable plates and record it below.

DATA AND CALCULATIONS

PLATE	A_1	A_2	B_1	B_2	C_1	C_2	D_1	D_2
Colonies counted								
Average # Colonies								

3. Calculate the original density (OD) using the following formula.

$$OD = \frac{\text{colonies counted}}{\text{FDF}}$$

Original density of *E. coli* in the broth (CFU/mL)	

REFERENCES

Collins, C. H., Patricia M. Lyne, J. M. Grange. 1995. Page 149 in *Collins and Lyne's Microbiological Methods*, 7th Ed. Butterworth-Heinemann, UK.
DIFCO Laboratories. 1984. Page 619 in *DIFCO Manual*, 10th Ed. DIFCO Laboratories, Detroit, MI.
Koch, Arthur L. 1994. Page 254 in *Methods for General and Molecular Bacteriology*, edited by Philipp Gerhardt, R. G. E. Murray, Willis A. Wood, and Noel R. Krieg, American Society for Microbiology, Washington, DC.
Postgate, J. R. 1969. Page 611 in *Methods in Microbiology*, Vol. 1., edited by J. R. Norris and D. W. Ribbons, Academic Press, Inc., New York.

6-2 URINE STREAK—SEMIQUANTITATIVE METHOD

Photographic Atlas Reference
Semiquantitative Streak of a Urine Specimen
Page 88

MATERIALS NEEDED FOR THIS EXERCISE

Per Student Group

- One blood agar plate (TSA with 5% sheep blood)
- One sterile volumetric inoculating loop (either 0.01 mL or 0.001 mL)
- A fresh urine sample

PROCEDURE

Lab One

1. Holding the loop vertically, immerse it in the urine sample. Then carefully withdraw it to obtain the correct volume of urine. This loop is designed to fill to capacity in the vertical position. Do not tilt it until you get it in position over the plate.
2. Inoculate the blood agar by making a single streak across the diameter of the plate.
3. Turn the plate 90° and, without flaming the loop, streak the urine across the entire surface of the agar as shown in Figure 6-3.
4. Invert, label, and incubate the plate for 24 hours at 37°C.

Lab Two

1. Remove the plate from the incubator and count the colonies. Also note any differing colony morphologies which would suggest possible colonization by more than one species.
2. Multiply the number of colonies by 100 if using a 0.01 mL loop; multiply by 1000 if using a 0.001 mL loop. (The number of cells contained in one milliliter is 100 times the number contained in 1/100th of a milliliter. Therefore, multiplying by the reciprocal of the volume gives the density in milliliters.) For a more detailed explanation, refer to *Photographic Atlas* page 88.
3. Enter the cell density in the table provided.

Figure 6-3 Semiquantitative Streak Method
Begin with a single streak across the center of the plate (I). Turn the plate 90° and do a zigzag streak (II) to spread the urine across the entire surface of the agar.

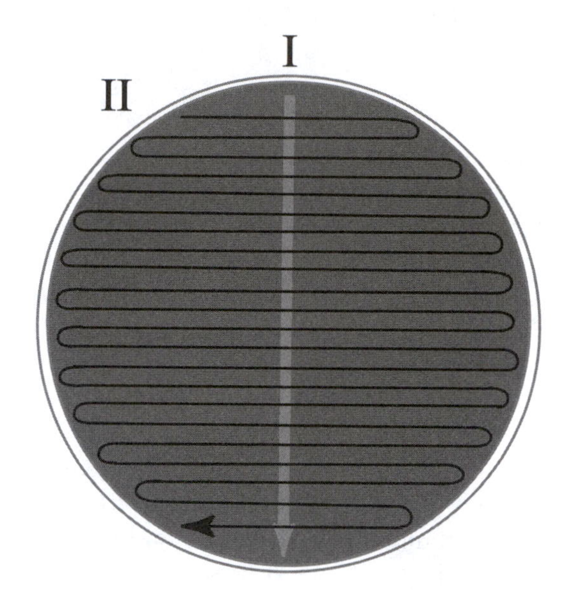

REFERENCES

Forbes, Betty A., Daniel F. Sahm, and Alice S. Weissfeld. 2002. Pages 933–934 in *Bailey and Scott's Diagnostic Microbiology*, 11th Ed. Mosby-Yearbook, St. Louis, MO.

Koneman, Elmer W., Stephen D. Allen, William M. Janda, Paul C. Schreckenberger, and Washington C. Winn, Jr. 1997. Page 94 in *Color Atlas and Textbook of Diagnostic Microbiology*, 5th Ed. J. B. Lippincott Company, Philadelphia, PA.

DATA AND CALCULATIONS

Enter you data in the table below. Multiply the number of colonies counted on the plate (A) with the reciprocal of your loop (B) to obtain the original cell density of the sample.

DATA AND CALCULATIONS

URINE SAMPLE	COLONIES COUNTED (A)	RECIPROCAL OF LOOP VOLUME (100 OR 1000) (B)	ORIGINAL CELL DENSITY (A X B)

6-3 DIRECT COUNT (PETROFF-HAUSSER)

Photographic Atlas Reference
Direct Count Page 85

MATERIALS NEEDED FOR THIS EXERCISE

Per Student Group

- Petroff-Hausser counting chamber with coverslip
- 1 mL pipettes with pipettor or micropipettes with 100 and 1000 µL tips
- Pasteur pipettes with bulbs or disposable transfer pipettes
- Hand counter
- Staining agents A and B (Appendix A)
- Test tubes for missing stain
- Overnight broth culture of:
  Proteus vulgaris

PROCEDURE

1. Transfer 100 µL (0.1 mL) from the original 24 hour culture tube to a non-sterile test tube.
2. Add 400 µL (0.4 mL) Agent A and 500 µL (0.5 mL) Agent B and mix well. [This dilution may not be suitable in all situations. There should be five to fifteen cells per small square for optimal results. Adjust the proportions of the broth culture and agents A and B if necessary to obtain a countable dilution, but remember to keep the total solution volume at 1000 µL (1.0 mL) for easier calculation of the dilution factor and cell density.]
3. Place a coverslip on the Petroff-Hausser counting chamber (Figure 6-4) and add a drop of the mixture at the edge of the coverslip. Capillary action will fill the well chamber.
4. Observe in the microscope and count the number of cells above at least 5 but no more than 16 small squares (Figure 6-5). Cells on a line belong to either the square below or to the right.
5. Enter your data in the space provided.
6. Calculate your dilution factor using the following formula.*

$$D_2 = \frac{(V_1)(D_1)}{V_2}$$

7. Calculate the original cell density of the *E. coli* culture using the following equation. (For a full explanation, refer to *Photographic Atlas* page 85.)

$$\text{Original cell density} = \frac{\text{Total cells counted}}{\left(\text{Squares counted}\right)\left(5 \times 10^{-8}\text{mL}\right)\left(\text{Dilution factor}\right)}$$

8. Record your answer in the table.

REFERENCES

Koch, Arthur L. 1994. Page 251 in *Methods for General and Molecular Bacteriology*, edited by Philipp Gerhardt, R. G. E. Murray, Willis A. Wood, and Noel R. Krieg, American Society for Microbiology, Washington, DC.

Postgate, J. R. 1969. Page 611 in *Methods in Microbiology*, Vol. 1., edited by J. R. Norris and D. W. Ribbons, Academic Press, Inc., New York.

* D_2 is the new dilution factor after the dilution takes place. V_1 is the volume of sample to be diluted. D_1 is the dilution factor of the sample before the dilution in question takes place (undiluted samples have a dilution factor of 1). V_2 is the combined volume of sample and diluent after the dilution takes place.

■ **FIGURE 6-4 Petroff-Hausser Counting Chamber**
The Petroff-Hausser counting chamber is a sophisticated microscope slide used for the direct counting of bacterial cells.

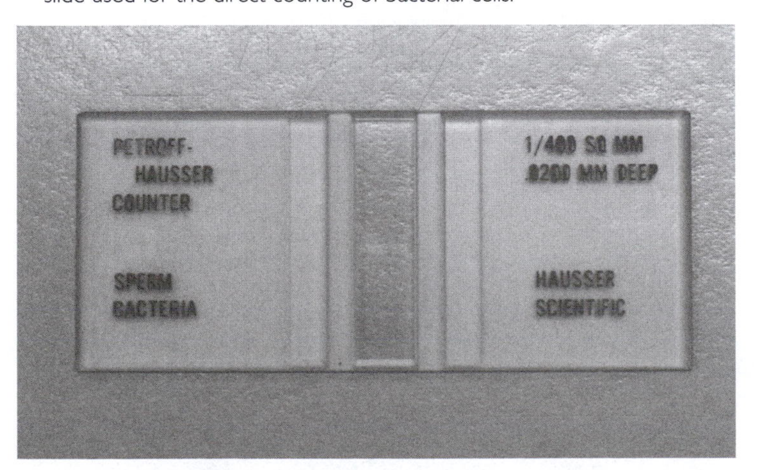

■ **FIGURE 6-5 Petroff-Hausser Counting Chamber Grid**
Shown is a portion of the Petroff-Hausser counting chamber grid. The smallest squares are the ones referred to in the formula. The volume above a small square is 5×10^{-8} mL (5×10^{-5} µL).

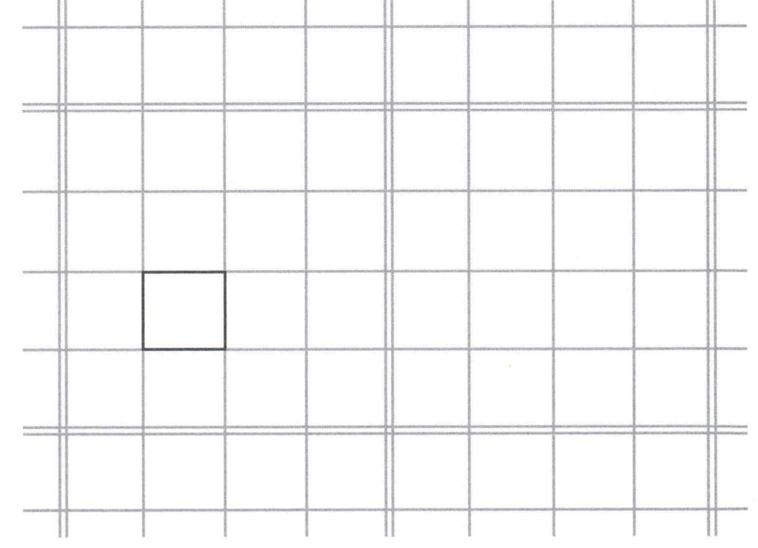

Data and Calculations

Enter you data in the table below. Calculate "dilution factor" and "original cell density" using the formula on page 133.

DATA AND CALCULATIONS			
Total Cells Counted	Squares Counted	Dilution Factor	Original Cell Density (calculated)

6-4 PLAQUE ASSAY

Photographic Atlas Reference
Plaque Assay Page 87

MATERIALS NEEDED FOR THIS EXERCISE

Per Student Group

- Micropipettes (10–100 µL and 100–1000 µL) with sterile tips
- Fourteen sterile capped microtubes (1.5 mL or larger)
- Seven nutrient agar plates
- Seven soft agar tubes
- Small flask of sterile water
- Hot water bath set at 45°C to keep the soft agar liquefied
- T4 coliphage
- 24 hour broth culture of *Escherichia coli B* (T-series phage host)

PROCEDURE

Refer to the Procedural Diagram in Figure 6-6 as needed.

Lab One

1. Obtain all materials except for the soft agar tubes. In order to keep the agar tubes liquefied, leave them in the water bath and take them out one at a time as needed.
2. Label seven microtubes 1 through 7. Label the other seven microtubes *E. coli* 1–7. Place all tubes in a rack pairing like-numbered tubes.
3. Label the Nutrient Agar plates A through G. Place them in the 35°C incubator to warm them. Take them out one at a time as needed. This will keep the soft agar (added at step 14) from solidifying too quickly and result in a smoother agar surface.
4. Aseptically transfer 990 µL sterile water to dilution tube 1.
5. Aseptically transfer 900 µL sterile water each to dilution tubes 2–7.

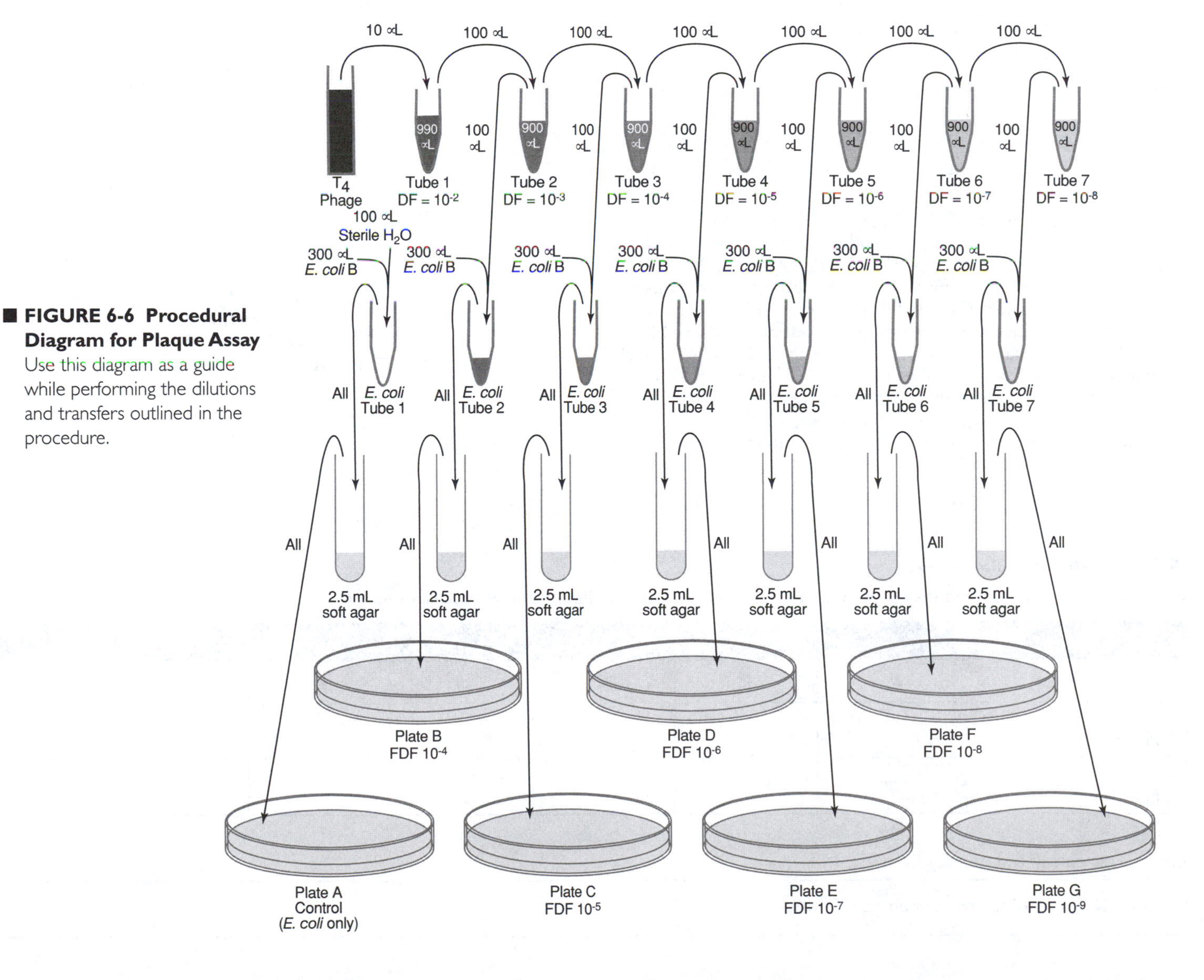

FIGURE 6-6 Procedural Diagram for Plaque Assay Use this diagram as a guide while performing the dilutions and transfers outlined in the procedure.

6. Mix the *E. coli* culture and aseptically transfer 300 µL into each of the *E. coli* microtubes.
7. Mix the T_4 suspension and aseptically transfer 10 µL to dilution tube 1. Mix well. This is DF (dilution factor) 10^{-2}.
8. Aseptically transfer 100 µL from dilution tube 1 to dilution tube 2. Mix well. This is DF 10^{-3}.
9. Aseptically transfer 100 µL from dilution tube 2 to dilution tube 3. Mix well. This is DF 10^{-4}.
10. Continue in this manner through dilution tube 7. The dilution factor of tube 7 should be 10^{-8}.
11. Aseptically transfer 100 µL sterile water to *E. coli* tube 1. This will be used to inoculate a control plate.
12. Aseptically transfer 100 µL from dilution tube 2 to its companion *E. coli* tube. Repeat this procedure with the remaining five tubes.
13. This is the beginning of the **preadsorption period**. Let all seven tubes stand undisturbed for 15 minutes.
14. Remove one soft agar tube from the hot water bath and add the contents of *E. coli* tube 1. Mix well and immediately pour onto plate A. Gently tilt the plate back and forth until the soft agar mixture is spread evenly across the solid medium. Label the plate "Control."
15. Remove a second soft agar tube from the water bath and add the contents of *E. coli* tube 2. Mix well and immediately pour onto plate B. Tilt back and forth to cover the agar and label it FDF* (final dilution factor) 10^{-4}.
16. Repeat this procedure with dilutions 10^{-4} thru 10^{-8} and plates C thru G. Label the plates with the appropriate FDF.
17. Allow the agar to solidify completely.
18. Invert the plates and incubate aerobically at 35°C for 24 to 48 hours.

* **Final Dilution Factors (FDF)**. Phage titer is calculated in PFU/mL based on the number of plaques produced on the plate. Plaque assay plates, however, are typically inoculated with only 0.1 mL (100 µL) of the diluted sample. If you do not account for this discrepancy the calculated original density will be off by a factor of ten. Therefore, since the plating of 0.1 mL dilution will yield 1/10th as many plaques as 1.0 mL would, the act of plating itself is treated as another tenfold dilution. So if the dilution factor of the source tube is 10^{-7}, the final dilution factor written on the plate will be 10^{-8}. In this way, the original density in PFU per milliliter can be simply determined by dividing the number of plaques by the FDF.

Lab Two

1. After incubation examine the control plate for growth and the absence of plaques.
2. Examine the remainder of your plates and determine which one is countable (30 to 300 plaques). Count the plaques and record the number in the table provided. Record all others as either TMTC (to many to count) or TFTC (to few to count) respectively.
3. Using the FDF on the countable plate and the following formula, calculate the original phage density. For more discussion on dilutions and formulas refer to *Photographic Atlas* pages 83 and 87.

$$\text{Original phage density (PFU/mL)} = \frac{\text{plaques counted}}{\text{FDF}}$$

4. Record your results in the space provided.

References

Collins, C. H., Patricia M. Lyne, J. M. Grange. 1995. Page 149 in *Collins and Lyne's Microbiological Methods*, 7th Ed. Butterworth-Heinemann, UK.

DIFCO Laboratories. 1984. Page 619 in *DIFCO Manual*, 10th Ed. DIFCO Laboratories, Detroit, MI.

Province, David L. and Roy Curtiss III. 1994. Page 328 in *Methods for General and Molecular Bacteriology*, edited by Philipp Gerhardt, R. G. E. Murray, Willis A. Wood and Noel R. Krieg, American Society for Microbiology, Washington, DC.

■ Data and Calculations

Record the number of plaques on your countable plate in the box below the appropriate letter.

DATA AND CALCULATIONS

Plate	A	B	C	D	E	F	G
Plaques counted							
FDF							

Enter your calculated density below.

Original density of the bacteriophage (PFU/mL)	

6-5 CLOSED SYSTEM GROWTH

Materials Needed For This Exercise

One of Each Per Student Group

- Five ice or water baths set at 10°C, 20°C, 25°C, 35°C, and 40°C
- Thermometers (to monitor actual temperatures in the ice and water baths)
- Five sterile side-arm flasks containing 49 mL of sterile Brain Heart Infusion (BHI) Broth + 2% NaCl
- Five sterile side-arm flasks containing 50 mL BHI + 2% NaCl to be used as controls
- Spectrophotometers
- Lab tissues
- Sterile 1 mL pipettes and pipettor or micropipettes (10–100 µL and 100–1000 µL) with sterile tips
- Bunsen burner
- Overnight broth culture of *Vibrio natriegens* in BHI + 2% NaCl (one per class)

Procedure

1. Collect all necessary materials. You will need a spectrophotometer, a sterile 1 mL pipette, and mechanical pipettor (or micropipette and tips), the side-arm flasks containing 49 mL and 50 mL of broth, and lab tissues. Label the 50 mL flask "control."
2. Immediately label your 49 mL growth flask; place it into the appropriate water bath and allow 15 minutes for the temperature to equilibrate. Turn on the spectrophotometer and let it warm up for a few minutes.
3. If the spectrophotometer is digital set it to "absorbance." Set the wavelength to 650 nm.
4. When all groups are ready to start, mix the *V. natriegens* culture and aseptically add 1.0 mL (1000 µL) to your growth flask. Mix it, wipe it dry making sure the side arm is clean, and immediately take a turbidity reading. This time is T_0. (**Caution:** Be careful not to do this too quickly; the side-arm flask is **VERY EASY TO SPILL** when it is tipped on its side for reading. Further, if it is pulled out on an angle, you risk breaking the side arm and spilling the culture.)
5. Record the absorbance next to your temperature under T_0 in the table provided. Remove the growth flask carefully and return it to the water bath.
6. Monitor temperature in your water/ice bath frequently. It is more important to have a constant temperature than to have exactly the assigned temperature. Record your *actual* temperature in Table 6-1.
7. Repeat the above procedure every 15 minutes (*i.e.*, T_{15}, T_{30}, T_{45}, etc.), taking turbidity readings and recording them in the table until you have completed and recorded the T_{240} reading. Make sure to place the flask in the spectrophotometer the same direction each time. Consistency in your readings will help compensate for irregularities in the glass.

Data and Calculations

1. Enter the absorbance values for all groups in Table 6-1 below.
2. On a computer or graph paper, plot your absorbance values versus time. This will produce a growth curve including growth phases as far as stationary phase. (Note: some cultures may not even reach stationary phase. Record what you get!). Also plot the growth curves of the other four samples on the same set of axes.
3. From your graph, determine the approximate length of time spent in the various growth stages for all samples and enter those values in Table 6-2.
4. From your graph, determine the absorbance for lag phase and stationary phase of each culture and enter those in Table 6-2.
5. Calculate the mean growth rate constant for each sample. Do this by choosing two points clearly on the linear part of exponential growth. These are A_1 and A_2. Determine

TABLE 6-1 ABSORBANCE READINGS

Temp (°C)	T_0	T_{15}	T_{30}	T_{45}	T_{60}	T_{75}	T_{90}	T_{105}	T_{120}	T_{135}	T_{150}	T_{165}	T_{180}	T_{195}	T_{210}	T_{225}	T_{240}
10																	
20																	
25																	
35																	
40																	

*These are recommended temperatures. Record the actual temperatures if they differ.

TABLE 6-2 GROWTH DATA

TEMP (°C)*	LAG PHASE (MINUTES)	EXPONENTIAL PHASE (MINUTES)	STATIONARY PHASE (MINUTES)	LAG PHASE (ABSORBANCE)	STATIONARY PHASE (ABSORBANCE)	A_1	A_2	k	g	t
10										
20										
25										
35										
40										

the absorbance values of each and the time in minutes (t) between the two points. Substitute your values in the equation and solve for (k). Enter your results in Table 6-2.

$$k = \frac{A_2 - A_1}{0.301t}$$

6. On a computer or graph paper, plot the mean growth rate versus temperature of the different samples.
7. Calculate generation times of the different samples. Enter your results in Table 6-2.

$$g = \frac{1}{k}$$

REFERENCES

Collins, C. H., Patricia M. Lyne, J. M. Grange. 1995. Page 149 in *Collins and Lyne's Microbiological Methods*, 7th Ed. Butterworth-Heinemann, UK.

Gerhardt, Philipp (Editor-in-chief), R. G. E. Murray, Willis A. Wood, and Noel R. Krieg (Editors). 1994. Chapter 11 in *Methods for General and Molecular Bacteriology*. American Society for Microbiology, Washington DC.

Koch, Arthur L. 1994. Page 251 in *Methods for General and Molecular Bacteriology*, edited by Philipp Gerhardt, R. G. E. Murray, Willis A. Wood, and Noel R. Krieg, American Society for Microbiology, Washington, DC.

Postgate, J. R. 1969. Page 611 in *Methods in Microbiology*, Vol. 1., edited by J. R. Norris and D. W. Ribbons, Academic Press, Inc., New York.

Prescott, Lansing M., John P. Harley and Donald A. Klein. 1999. Page 114 in *Microbiology*, 4th Ed. WCB/McGraw-Hill, Boston, MA.

SECTION SEVEN

7 Environmental, Food, and Medical Microbiology

The three microbiological disciplines examined in this section have distinctly different scopes of practice while overlapping to a significant degree in the public health domain. Medical microbiology is the study and control of pathogenic and potentially pathogenic microorganisms, the infections they cause and the host factors that interact with them. Environmental microbiology is the study, utilization, and control of microorganisms in the environment. Many environmental organisms are used beneficially by industry such as in the production of vinegar, leaching of low-grade ores, or sewage purification. Other environmental organisms are of human or animal origin (usually fecal) and threaten us with reintroduction as potential contaminants of food or water. Food microbiology is devoted to the study and utilization of beneficial microbes as well as the control of many common yet potentially deadly contaminants. Many molds, yeasts, and bacteria are responsible for production, preservation, and flavoring of foods we love to eat while others produce toxins so powerful that as little as one nanogram per kilogram of body weight can be lethal.

So to a certain degree, all microbiological disciplines are medically oriented. Circumstances that encourage growth of beneficial organisms also encourage deadly ones. The fact that microorganisms proliferate under the same conditions necessary for other life forms makes them competitors for limited resources. And, perhaps most importantly, microbial ability to adapt and become resistant to traditional control measures makes them a constant and persistent challenge for all microbiologists.

Environmental Microbiology

Environmental microbiology is the study, utilization, and control of microorganisms living in marine, freshwater, or terrestrial habitats. In efforts to keep our environment safe from disease-causing contamination, local and national governmental regulatory agencies perform daily tests similar to those introduced in this unit. The first two tests—the membrane filter technique and the most probable number technique—are used to measure fecal contamination of water. The microbial soil count is used to isolate and demonstrate the variety of populations residing in soil. Bioluminescence, our final exercise in this unit, is the light-emitting phenomenon expressed by certain marine bacteria.

7-1 MEMBRANE FILTER TECHNIQUE

Photographic Atlas Reference
Membrane Filter Technique Page 92

Materials Needed For This Exercise

Per Student Group

- One Endo Agar plate
- One sterile membrane filter (pore size 0.45 μm)
- Sterile membrane filter suction apparatus (Figure 7-1)
- 100 mL water sample (obtained by student)
- Gloves
- Household disinfectant and paper towels
- Small beaker containing alcohol and forceps
- Vacuum source (pump or aspirator)

Procedure

Prelab

1. Obtain a 100 mL water dilution bottle from your instructor.
2. Choose an environmental source to sample. (Your instructor may decide to approve your choice to avoid duplication among lab groups.)
3. Visit the environmental site as close to your lab period as possible. Bring your water dilution bottle, a pair of gloves, some household disinfectant, and paper towels. While wearing gloves, fill the bottle to the white line (100 mL) and replace the cap.
4. Wipe the outside of the bottle with disinfectant. Dispose of the towels and gloves in the trash.
5. Store the sample in the refrigerator until your lab period. If the sample must sit for awhile before your lab, leave the cap loose to allow some aeration. Be careful not to spill!

Lab One

1. Alcohol-flame the forceps and place the membrane filter (grid facing up) between the two halves of the filter housing (Figure 7-2). Clamp the two halves of the filter housing together.

FIGURE 7-1 Membrane Filter Suction Apparatus
Assemble the membrane filter apparatus as shown in this photograph. It is important to use two suction flasks (as shown) to avoid getting water into the vacuum source. Secure the flasks on the table as the tubing may make them top heavy and unstable.

FIGURE 7-2 Place the Filter on the Filter Housing
Carefully place the filter on the bottom half of the filter housing. Clamp the filter funnel over the filter.

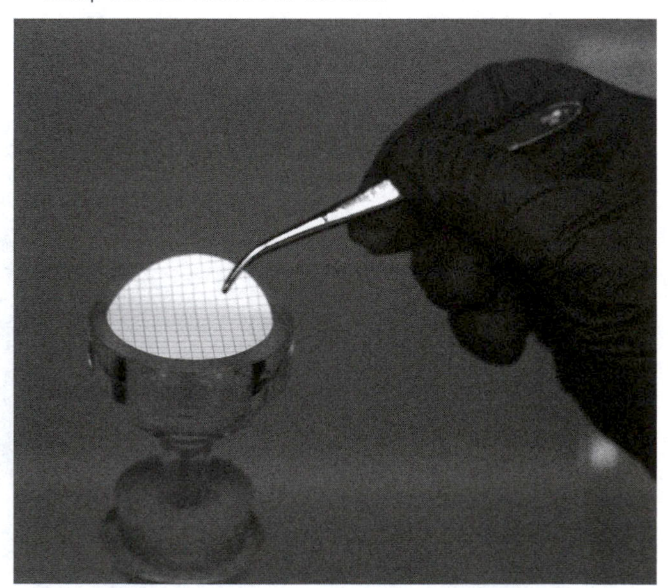

2. Insert the filter housing into the suction flask as shown in Figure 7-3. (This assembly can be a little top heavy; have someone hold it or otherwise secure it to prevent tipping.)
3. Pour the appropriate volume of water sample into the funnel. (Refer to Table 7-1 for suggested sample volumes. If the sample size is smaller than 10.0 mL add 10 to 20 mL of sterile water before filtering. This will help distribute the cells evenly on the surface of the membrane filter.)
4. Turn on the suction pump (or aspirator) and filter the sample into the flask.
5. Sterilize the forceps again and carefully transfer the filter to the Endo agar plate, being careful not to fold it or create air pockets (Figure 7-4).

■ FIGURE 7-3 Membrane Filter Assembly
The membrane filter assembly is made up of a two-piece funnel and clamp. The membrane filter is inserted between the two funnel halves and the whole assembly is clamped together.

■ FIGURE 7-4 Place the Filter on the Agar Plate
Using sterile forceps carefully place the filter onto the agar surface with the grid facing up. Try not to allow any air pockets under the filter since contact with the agar surface is essential for bacterial growth. Allow a few minutes for the filter to adhere to the agar before inverting the plate.

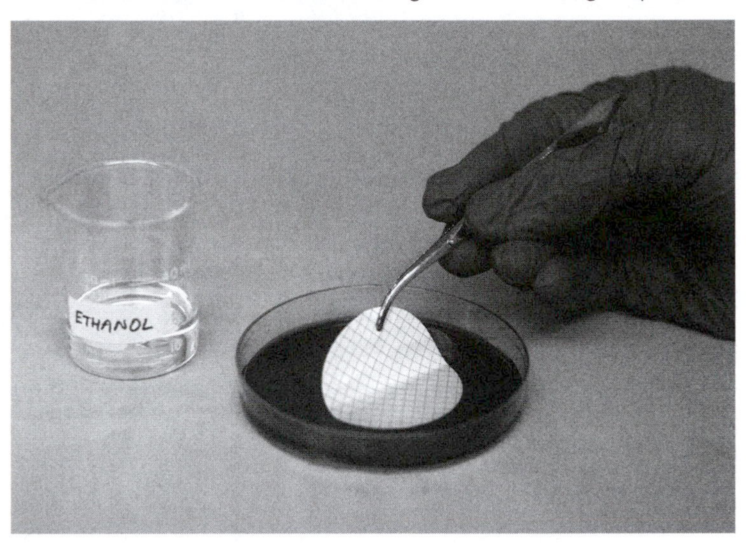

6. Wait a few minutes to allow the filter to adhere to the agar, then invert the plate and incubate it aerobically at 35°C for 48 hours.

Lab Two

1. Remove the plate and count the colonies on the membrane filter that are dark purple, have a black center, or produce a green metallic sheen.
2. Record your data in the table provided.
3. Calculate the coliform CFU per 100 milliliters using the following formula:

$$\text{Coliforms/100 mL} = \frac{\text{(coliform colonies counted) (100)}}{\text{(mL of original sample filtered)}}$$

4. Record your results in the table.

TABLE 7-1 SUGGESTED SAMPLE VOLUMES FOR MEMBRANE FILTER TEST (TABLE COURTESY OF AMERICAN PUBLIC HEALTH ASSOCIATION)

Source	Volume to be Filtered (mL)							
	100	50	10	1.0	0.1	0.01	0.001	0.0001
Drinking water	X							
Swimming Pool	X							
Lake	X	X	X					
Well water	X	X	X					
Public Beach			X	X	X			
River				X	X	X	X	
Raw Sewage					X	X	X	X

DATA AND CALCULATIONS

Enter the number of colonies counted in each sample and calculate the total coliforms per 100 milliliters of water.

DATA AND CALCULATIONS		
SAMPLE	**NUMBER OF COLONIES**	**COLONIES/100 mL**

REFERENCES

Chan, E. C. S., Pelczar, Jr., Noel R. Krieg 1986. Page 291 in *Laboratory Exercises In Microbiology*. McGraw-Hill Book Company.

Clesceri, WEF, Chair; Arnold E. Greenberg, APHA; Andrew D. Eaton, AWWA; and Mary Ann H. Franson. 1998. Chapter 9 in *Standard Methods for the Examination of Water and Wastewater*, 20th Edition. American Public Health Association, American Water Works Association, Water Environment Federation. APHA Publication Office, Washington, DC.

Collins, C. H., Patricia M. Lyne, J. M. Grange. 1995. Page 270 in *Collins and Lyne's Microbiological Methods*, 7th Ed. Butterworth-Heinemann, UK.

DIFCO Laboratories. 1984. Page 515 in *DIFCO Manual*, 10th Ed. DIFCO Laboratories, Detroit, MI.

Mulvany, J. G. 1969. Page 205 in *Methods in Microbiology*, Vol. 1, edited by J. R. Norris and D. W. Ribbons, Academic Press Inc., New York.

Power, David A. and Peggy J. McCuen. 1988. Page 153 in *Manual of BBL® Products and Laboratory Procedures*, 6th Ed. Becton Dickinson Microbiology Systems, Cockeysville, MD.

7-2 MULTIPLE TUBE FERMENTATION METHOD FOR TOTAL COLIFORM DETERMINATION OR MOST PROBABLE NUMBER (MPN) METHOD

Photographic Atlas Reference
MPN Page 93

MATERIALS NEEDED FOR THIS EXERCISE

Per Student Group

- Fifteen lauryl tryptose broth (LTB) tubes (containing 10 mL broth and an inverted Durham tube)
- Up to fifteen brilliant green lactose bile (BGLB) broth tubes (The number of tubes required will be determined by the results of the LTB test.)
- Up to fifteen EC broth tubes (The number of tubes required will be determined by the results of the LTB test.)
- Water sample (May be obtained by student)
- Two 9.0 mL dilution tubes
- Sterile 1.0 mL pipettes and pipettor
- Water bath set at 45.5°C
- Test tube rack
- Labeling tape

PROCEDURE

Lab One

1. Arrange the 15 LTB tubes into three groups of five in a test tube rack. Label the groups A, B, and C respectively.
2. Aseptically transfer one mL of the (undiluted) water sample to each LTB tube in the group labeled A. Mix well. This is undiluted sample so its dilution factor (DF) is 10^0. ($10^0 = 1$; *i.e.*, no dilution.)
3. Make a dilution by adding 1.0 mL of the water sample to one of the 9.0 mL dilution tubes. This dilution contains only 1/10 mL original sample to 9/10 mL sterile water; therefore, the DF is 1/10 or 0.1 or 10^{-1} (preferred). For help with dilutions and dilution factors, refer to Exercise 6-1 in this manual and supporting text in the *Photographic Atlas*, page 83.
4. Add 1.0 mL of the 10^{-1} dilution (containing 0.1 mL of the original sample) to each of the LTB tubes in group B. Mix well.
5. Make a second dilution by adding 1.0 mL of the 10^{-1} dilution to one of the 9.0 mL dilution tubes. Mix well. This dilution contains 1/100 mL original sample to 99/100 mL sterile water and has a DF of 10^{-2}.
6. Add 1.0 mL of the 10^{-2} dilution (containing 0.01 mL of the original sample) to each of the LTB tubes in the C group.
7. Incubate the LTB tubes at 35 to 37°C for 48 hours.

Lab Two

1. Remove the broths from the incubator and, one group at a time, examine the Durham tubes for accumulation of gas. Gas production is a positive result; absence of gas is negative. Record positive and negative results in the table provided. Refer to the example in Table 7-2.
2. Using an inoculating loop, inoculate one BGLB broth with each positive LTB tube showing evidence of gas production. (Make sure each BGLB tube is *clearly labeled* A, B, or C according to the LTB tube from which it is inoculated.)
3. Inoculate EC broths with the positive LTB tubes in the same manner as the BGLB above. Again be sure to clearly label all EC tubes appropriately.
4. Incubate the BGLB at 35 to 37°C for 48 hours. Incubate the EC tubes in the 45.5°C water bath for 48 hours.

Lab Three

1. Remove all tubes from the incubator and water bath and examine the Durham tubes for gas accumulation. Count the positive BGLB tubes and enter your results in the table provided.
2. Using Table 7-2 as a guide complete the BGLB Data table.
3. Using the data in the BGLB Data table calculate the total coliform MPN using the following formula. (This formula is used to calculate both Total Coliform MPN and *E. coli* MPN.)

$$\text{MPN/100 mL} = \frac{(\text{Total number of positive results})\ (100)}{\sqrt{\begin{pmatrix}\text{Combined volume of}\\ \text{sample in negative tubes}\end{pmatrix}\begin{pmatrix}\text{Combined volume of}\\ \text{sample in all tubes}\end{pmatrix}}}$$

4. Count the positive EC broth results in the same manner as the BGLB test. Record the results in the table provided.
5. Determine the *E. coli* MPN in the same manner as described for total coliform count and record your results in the table.

EXAMPLE: Substituting the values from Table 7-2, calculation of total coliform MPN for the water sample in the example is as follows:

$$\text{MPN/100 mL} = \frac{(\text{Total number of positive results})\ (100)}{\sqrt{\begin{pmatrix}\text{Combined volume of}\\ \text{sample in negative tubes}\end{pmatrix}\begin{pmatrix}\text{Combined volume of}\\ \text{sample in all tubes}\end{pmatrix}}}$$

$$\text{MPN/100 mL} = \frac{9 \times 100}{\sqrt{(.24 \times 5.55)}} = \frac{900}{\sqrt{1.332}} = \frac{900}{1.154} = 780$$

$$\text{MPN} = 780 \text{ coliforms/100 mL}$$

Reference

Clesceri, Lenore S., WEF, Arnold E. Greenberg, APHA, and Andrew D. Eaton, AWWA. 1998. *Standard Methods for the Examination of Water and Wastewater*, 20th Ed. Prepared and published jointly by American Public Health Association, American Water Works Association, and Water Environment Federation. APHA publication office, Washington, DC.

Use this example when completing the tables and calculating the necessary components (shaded area) to insert into the MPN equation. In this hypothetical test, 15 tubes were tested in groups of 5. The sample was used in three dilutions (10^0, 10^{-1}, and 10^{-2}) to inoculate the broths. The second row contains the dilution factor of inoculum used per group. The third row shows the actual amount of original sample that went into each broth. The fourth row contains the number of tubes in each group (5 in this example). The fifth row shows the number of tubes from each group of 5 that showed evidence of gas production. The sixth row shows the number of tubes from each group that did *not* show evidence of gas production. The seventh row is used for calculating the "combined volume of sample in negative tubes" and refers to the inoculum that went into the LTB tubes. This total inserts into the equation. The eighth row is used for calculating the "combined volume of sample in all tubes" and refers to the inoculum that went into the LTB tubes. This total inserts into the equation. As you can see the undiluted inoculation produced 5 positive results and 0 negative results, the 10^{-1} dilution produced 3 positive results and 2 negative, and the 10^{-2} dilution produced 1 positive and 4 negative results. The total volume of original sample that went into tubes was 5.55 mL, 0.24 mL of which produced no gas.

TABLE 7-2 RESULTS OF A HYPOTHETICAL BGLB TEST

GROUP	A	B	C	TOTALS (A + B + C)
Dilution Factor (DF)	10^0	10^{-1}	10^{-2}	NA
Portion of dilution added to LTB tubes that is original sample (1.0 ML X DF)	1.0 mL	0.1 mL	0.01 mL	NA
# Tubes in group	5	5	5	NA
# Positive results (gas)	5	3	1	9
# Negative results (no gas)	0	2	4	NA
Volume of original sample in negative LTB tubes (DF X 1.0ML X # negative tubes)	0 mL	0.2 mL	0.04 mL	0.24 mL
Volume of original sample in all LTB tubes (DF X 1.0ML X # tubes)	5.0 mL	0.5 mL	0.05 mL	5.55 mL

■ DATA AND CALCULATIONS

Enter your data here.

DATA AND CALCULATIONS BGLB DATA				
GROUP	**A**	**B**	**C**	**TOTALS (A + B + C)**
Dilution Factor (DF)	10^0	10^{-1}	10^{-2}	NA
Portion of dilution added to LTB tubes that is original sample (1.0ML X DF)	1.0 mL	0.1 mL	0.01 mL	NA
# Tubes in group	5	5	5	NA
# Positive results (gas)				
# Negative results (no gas)				NA
Volume of original sample in negative LTB tubes (DF X 1.0ML X # negative tubes)				
Volume of original sample in all LTB tubes (DF X 1.0ML X # tubes)	5.0 mL	0.5 mL	0.05 mL	5.55 mL

Enter your final results here.

DATA AND CALCULATIONS EC DATA				
GROUP	**A**	**B**	**C**	**TOTALS (A + B + C)**
Dilution Factor (DF)	10^0	10^{-1}	10^{-2}	NA
Portion of dilution added to LTB tubes that is original sample (1.0ML X DF)	1.0 mL	0.1 mL	0.01 mL	NA
# Tubes in group	5	5	5	NA
# Positive results (gas)				
# Negative results (no gas)				
Volume of original sample in negative LTB tubes (DF X 1.0ML X # negative tubes)				
Volume of original sample in all LTB tubes (DF X 1.0ML X # tubes)	5.0 mL	0.5 mL	0.05 mL	5.55 mL

Total coliform MPN/100 mL (from BGLB data)	
E. coli MPN/100 mL (from EC data)	

7-3 BIOLUMINESCENCE

Photographic Atlas Reference
Bioluminescence Page 95

A few marine bacteria from genera *Vibrio* and *Photobacterium* are able to emit light by a process known as **bioluminescence**. Many of these organisms maintain mutualistic relationships with other marine life. For example, *Photobacterium* species living in an animal called the Flashlight Fish receive nutrients from the fish and in return provide a unique device for frightening would-be predators.

Bioluminescent bacteria are given the ability to emit light by an enzyme called **luciferase**. In the presence of oxygen and a long-chain aldehyde (R–CHO) luciferase catalyzes the oxidation of reduced flavin mononucleotide ($FMNH_2$). In the process, outer electrons surrounding FMN become excited. Light is emitted when the electronically excited FMN returns to its ground state.

$$FMNH_2 + O_2 + RCHO \xrightarrow{\text{luciferase}} FMN + RCOOH + H_2O + \text{light}$$

In this exercise you will inoculate Photobacterium Broth with *Vibrio fischeri*. Because the best results are produced in the presence of abundant oxygen, incubation will take place in a warm shaker water bath.

Materials Needed For This Exercise

Per Student Group

- Small flasks of sterile Photobacterium Broth
- Shaker water bath set at 30°C
- Overnight culture of *Vibrio fischeri*

Procedure

Lab One

1. Obtain a culture of *Vibrio fischeri* and aseptically transfer 1.0 mL into a growth flask.
2. Place the flask into the shaker water bath; start the shaker and incubate for 24 to 48 hours.

Lab Two

1. Remove the flask from the water bath and examine it in a dark room for light emission. It may take awhile for your eyes to adjust to the dark and see the bioluminescence.

References

DIFCO Laboratories. 1984. *DIFCO Manual*, 10th Ed. DIFCO Laboratories, Detroit, MI.

Krieg, Noel R. and John G. Holt (Editor-in-Chief). 1984. Page 518 in *Bergey's Manual of Systematic Bacteriology*, Vol. 1. Lippincott Williams and Wilkins, Baltimore, MD.

Power, David A. and Peggy J. McCuen. 1988. *Manual of BBL® Products and Laboratory Procedures*, 6th Ed. Becton Dickinson Microbiology Systems, Cockeysville, MD.

7-4 SOIL MICROBIAL COUNT

This exercise is designed to give you an idea of the extraordinary diversity and volume of cells that can be found in even a small sample of soil. In today's lab you will perform a serial dilution (Exercise 6-1) of a soil sample and calculate the densities of three very prominent soil residents—bacteria, actinomycetes, and fungi. Colony counts will be performed on plates produced using the pour plate technique described in Exercise 6-4.

The media used for today's exercise are specifically designed for the organisms you are trying to isolate. Glycerol Yeast Extract Agar is designed for actinomycetes and doesn't offer enough nutritive value for typical bacteria or fungi. Nutrient Agar is designed for a broad range of bacteria, but not fungi. Sabouraud Dextrose Agar is designed for fungi but will support bacteria, therefore penicillin and streptomycin have been added to discourage bacterial growth.

To save time and material, the class will be divided into six groups, as follows:

GROUP ASSIGNMENTS

Group Number	Organism	Media (see materials)
1	Actinomycetes	GYEA
2	Actinomycetes	GYEA
3	Bacteria	NA
4	Bacteria	NA
5	Fungi	SDA
6	Fungi	SDA

Materials Needed For This Exercise

Per Student Group

- Soil sample
- Water bath set at 45°C containing sterile molten:
  - Sabouraud Dextrose Agar
  - Nutrient Agar
  - Glycerol Yeast Extract Agar
- Capped bottle containing 90 mL sterile water
- Micropipettes (10–100 µL and 100–1000 µL) with sterile tips
- Sterile microtubes
- Flask of sterile water
- 5 Sterile Petri dishes
- Hand tally counter
- Colony counter

Procedure

Lab One—Groups 1 and 2

1. Obtain all materials including five Petri dishes. Label the plates GYEA—A through E respectively. Plate E will be the control.
2. Obtain 5 microtubes and label them 1 through 5. These are your dilution tubes; make sure they remain covered until needed.
3. Aseptically add 900 µL (0.9 mL) sterile water to dilution tubes 2, 3, 4, and 5. Cover when finished.
4. Add 10 g of the soil sample to the 90 mL water bottle. Shake vigorously for several minutes. This is DF (dilution factor) 10^{-1}. If you don't remember how to calculate DF, refer to Exercise 6-1.
5. Aseptically transfer 1000 µL of the suspended solution from the bottle to dilution tube 1. Remember, this is DF 10^{-1}
6. Aseptically transfer 100 µL from dilution tube 1 to dilution tube 2; mix well. This is DF 10^{-2}.
7. Aseptically transfer 100 µL from dilution tube 2 to dilution tube 3; mix well. This is DF 10^{-3}.
8. Aseptically transfer 100 µL from dilution tube 3 to dilution tube 4; mix well. This is DF 10^{-4}.
9. Tube 5 will remain uninoculated as a control.
10. Remove one molten GYEA tube from the water bath. Aseptically add 100 µL from dilution tube 1, mix well and pour into plate A. Repeat the process using 100 µL from tubes 2 through 5 to plates B through E respectively. Label the plates with the FDF (see footnote, page 136). Label plate E "control."
11. Allow the plates time to cool and solidify. Invert and incubate them at 25°C for two to seven days.

Lab One—Groups 3 and 4

1. Obtain all materials including five Petri dishes. Label the plates NA—A through E respectively. Plate E will be the control.
2. Obtain 5 microtubes and label them 1 through 5. These are your dilution tubes; make sure they remain covered until needed.
3. Aseptically add 900 µL sterile water to each dilution tube. Cover when finished.
4. Group 1 will have added 10 g of the soil sample to the 90 mL water bottle. (This is DF 10^{-1}. If you don't remember how to calculate dilution factors, refer to Exercise 6-1.)
5. Obtain the bottle from Group 1, mix the sample well and aseptically transfer 100 µL of the suspended solution to dilution tube 1. Mix the contents of tube 1. This is DF 10^{-2}.
6. Aseptically transfer 100 µL from dilution tube 1 to dilution tube 2; mix well. This is DF 10^{-3}.
7. Aseptically transfer 100 µL from dilution tube 2 to dilution tube 3; mix well. This is DF 10^{-4}.

8. Aseptically transfer 100 µL from dilution tube 3 to dilution tube 4; mix well. This is DF 10^{-5}.
9. Tube 5 will remain uninoculated as a control.
10. Remove one molten NA tube from the water bath. Aseptically add 100 µL from dilution tube 1, mix well and pour into plate A. Repeat the process using 100 µL from tubes 2 through 5 to plates B through E respectively. Label the plates with the FDF (See footnote on Page 136). Label plate E "control."
11. Allow the plates time to cool and solidify. Invert and incubate them at 25°C for two to seven days.

Lab One—Groups 5 and 6

1. Obtain all materials including five Petri dishes. Label five plates SDA—A through E respectively. Plate E will be the control.
2. Obtain 4 microtubes and label them 1 through 4. These are your dilution tubes; make sure they remain covered until needed.
3. Aseptically add 900 µL (0.9 mL) sterile water to dilution tubes 2 through 4. Cover when finished.
4. Group 1 will have added 10 g of the soil sample to the 90 mL water bottle. (This is DF 10^{-1}. If you don't remember how to calculate dilution factors, refer to Exercise 6-1.)
5. Obtain the bottle containing the sample, mix well, and aseptically transfer 1500 µL of the suspended solution to dilution tube 1. Remember, this is DF 10^{-1}.
6. Aseptically transfer 100 µL from dilution tube 1 to dilution tube 2; mix well. This is DF 10^{-2}.
7. Aseptically transfer 100 µL from dilution tube 2 to dilution tube 3; mix well. This is DF 10^{-3}.
8. Tube 4 will remain uninoculated as a control.
9. Remove one molten SDA tube from the water bath. Aseptically add 1000 µL (1.0 mL) from dilution tube 1; mix well, and pour into plate A. Aseptically add 100 µL from tube 1 to an agar tube; mix well, and pour into plate B. Repeat the process using 100 µL from tubes 2 through 4 to plates C through E respectively. Label the plates with the FDF (the final dilution factor of plate A will remain the same as the source tube because it was inoculated with 1000 µL; see footnote on Page 136). Label plate E "control."
10. Allow the plates time to cool and solidify. Invert and incubate them at 25°C for two to seven days.

Lab Two—All Groups

1. After incubation set the countable plates aside (plates with 30 to 300 colonies) and properly dispose of the uncountable plates. Only one plate should be countable.
2. Count the colonies on the plate and calculate the cell density of the water using the following formula.

$$\text{Original density (CFU/mL)} = \frac{\text{Colonies counted}}{\text{FDF}}$$

3. Since your task is to determine the original density per gram of soil you must convert cell density in milliliters to grams as follows (remembering that the original 100 milliliters of solution contained 10 grams of soil):

$$\text{Original density (CFU/mL)} \times \frac{100\text{ mL}}{10\text{ g}} = \text{Original density (CFU/g)}$$

REFERENCES

Benson, Harold J. 1998. Page 196 in *Microbiological Applications, Laboratory Manual in General Microbiology*, 7th Edition. WCB McGraw-Hill, Boston, MA.

DIFCO Laboratories. 1984. Page 768 in *DIFCO Manual*, 10th Ed. DIFCO Laboratories, Detroit, MI.

Krieg, Noel R. and John G. Holt (Editor-in-Chief). 1984. Page 1383 in *Bergey's Manual of Systematic Bacteriology*, Vol. 1. Lippincott Williams and Wilkins, Baltimore, MD.

Varnam, Alan H. and Malcolm G. Evans. 2000. Page 88 in *Environmental Microbiology*. ASM Press, Washington, DC.

■ DATA AND CALCULATIONS

Count the colonies on the countable plates and enter the data below.

DATA AND CALCULATIONS

GROUP	COLONIES COUNTED	FINAL DILUTION FACTOR (FDF) OF PLATED SAMPLE	CELL DENSITY PER ML OF H_2O	CELL DENSITY PER GRAM OF SOIL
1				
2				
3				
4				
5				
6				

Food Microbiology

Food microbiology is the study, utilization, and control of microorganisms in food. Many of our favorite foods such as yogurt, wine, beer, sauerkraut, buttermilk, vinegar, bread, and cheeses are produced by or with the help of microorganisms. On the other hand, some of the most potent toxins can be found in under-processed canned foods. Many illnesses are caused by and can be prevented by avoiding improper handling of foods. In this section you will use beneficial organisms to produce a food, you will perform a test of milk quality, and you will test the effectiveness of a typical fermentation starter culture as a food preservative.

7-5 METHYLENE BLUE REDUCTASE TEST

Photographic Atlas Reference
Methylene Blue Reductase Test Page 95

MATERIALS NEEDED FOR THIS EXERCISE

Per Student Group

- Raw milk samples
- Sterile screw-capped test tubes
- Sterile 1 mL and 10 mL pipettes
- Hot water bath set at 35°C
- Methylene blue reductase reagent
- Overnight broth culture of *Escherichia coli*
- Clock or wristwatch

PROCEDURE

Lab One

1. Obtain sterile tubes for as many samples as you are testing plus two more to be used as positive and negative controls. Label them appropriately.
2. Using a sterile 10 mL pipette, aseptically add 10 mL milk to each tube.
3. Inoculate the tube marked "positive control" with 1 mL of *E. coli* culture.
4. Aseptically add 1.0 mL methylene blue solution to each test tube. Cap the tubes tightly and invert them several times to mix thoroughly.
5. Place the tube marked "negative control" in the refrigerator to prevent it changing color.
6. Place all other tubes in the hot water bath and note the time.
7. After 5 minutes remove the tubes, invert them once to mix again then return them to the water bath. Record the time in the table under "Starting Time."
8. Using the control tubes for color comparison, check tubes at 30-minute intervals and record the time when each becomes white. Poor quality milk takes less than 2 hours; good quality milk takes longer than 6 hours. If necessary have someone check the tubes at 6 hours and record the results for you.
9. Using the table provided, calculate the time it takes for each milk sample to become white.

REFERENCES

Bailey, R. W., and E. G. Scott. 1966. Page 114 and 306 in *Diagnostic Microbiology*, 2nd Ed. C. V. Mosby Company, St. Louis, MO.

Benathen, Isaiah. 1993. Page 132 in *Microbiology With Health Care Applications*. Star Publishing Company, Belmont, CA.

Power, David A. and Peggy J. McCuen. 1988. Page 62 in *Manual of BBL® Products and Laboratory Procedures*, 6th Ed. Becton Dickinson Microbiology Systems, Cockeysville, MD.

Richardson (ed.). 1985. *Standard Methods for the Examination of Dairy Products*, 15th Ed. American Public Health Association, Washington DC.

DATA AND CALCULATIONS

Enter data below and determine the quality of your milk. Under "Milk Quality," record "G" (good) if the milk takes longer than six hours to turn white, "P" (poor) if it turns white in two hours or less, and "M" (medium quality) if it turns white between two and six hours after inoculation.

DATA AND CALCULATIONS

SAMPLE	STARTING TIME T_S (MILK IS BLUE)	ENDING TIME T_E (MILK IS WHITE)	ELAPSED TIME $(T_E - T_S)$	MILK QUALITY

7-6 VIABLE CELL PRESERVATIVES

Lactococcus lactis, (formerly called *Streptococcus lactis*), is a naturally occurring organism in milk that produces the antibiotic **nisin.** It is believed that nisin, from *L. lactis* (even when not actively metabolizing) slows growth of refrigerated psychotrophic organisms. In this exercise you will compare the times required for *Enterococcus*, *Staphylococcus*, *Pseudomonas*, *Clostridium*, and *Salmonella* organisms to curdle treated and untreated refrigerated milk. This is a long-term exercise with readings performed semiweekly up to four weeks.

MATERIALS NEEDED FOR THIS EXERCISE

Per Student Group

- Sterile skim milk medium
- Sterile 1 ml pipettes
- Methylene blue
- Pipette pump
- Thermometer
- Overnight broth cultures of:
  - *Pseudomonas fluorescens*
  - *Clostridium sporogenes*
  - *Staphylococcus aureus*
  - *Salmonella typhimurium*
  - *Enterococcus faecalis*
  - *Lactococcus lactis*

PROCEDURE

Lab One

1. Obtain twelve skim milk tubes and one of each culture. Organize the tubes into pairs and label them by name, two for each organism. Label one of each pair "L" to indicate it is a tube to receive *L. lactis*. Label the two extra tubes "control" and "control L."
2. Mix the cultures well.
3. Add three drops of each culture to its correspondingly labeled tubes.
4. Add one mL of *L. lactis* to each milk broth labeled with an "L" including "control L." Mix.
5. Immediately place all tubes in a refrigerator set at 5°C to 10°C.

Lab Two

1. Observe the skim milk medium for curdling twice per week for up to four weeks.
2. Record your results in the table provided.
3. Optional: On a sheet of graph paper, prepare a histogram of organism vs. time. Be sure to include the control. It may be helpful to construct double bars comparing treated samples with untreated samples.

REFERENCES

Krieg, Noel R. and John G. Holt (Editor-in-Chief). 1984. Page 518 in *Bergey's Manual of Systematic Bacteriology*, Vol. 1. Lippincott Williams and Wilkins, Baltimore, MD.

Ray, Bibek. 2001. Chapter 13 in *Fundamental Food Microbiology*, 2nd Ed. CRC Press LLC, Boca Raton, FL.

■ OBSERVATIONS

Enter "C" for curdled; "NC" for not curdled.

OBSERVATIONS													
TIME	CONTROL		*P. FLUORESCENS*		*C. SPOROGENES*		*E. FAECALIS*		*S. AUREUS*		*S. TYPHIMURIUM*		
	TREATED	UNTREATED	TREATED	UNTREATED	TREATED	UNTREATED	TREATED	UNTREATED	TREATED	UNTREATED	TREATED	UNTREATED	
.5 wk													
1 wk													
1.5 wks													
2 wks													
2.5 wks													
3 wks													
3.5 wks													
4 wks													

7-7 MAKING YOGURT

Several species of bacteria are used in the commercial production of yogurt. Most formulations include combinations of two or more species to synergistically enhance growth and to produce the optimum balance in flavor and acidity. One common pairing of organisms in commercial yogurt is that of *Lactobacillus delbrueckii* subsp *bulgaricus* and *Streptococcus thermophilus*.

Yogurt gets its unique flavor from acetaldehyde, diacetyl, and acetate produced from the fermentation of the milk sugar lactose. The proportions of products, and ultimately the flavor, in the yogurt depend upon the types of enzyme systems possessed by the species used. Both species mentioned above contain **constitutive** β-galactosidase systems that break down lactose and convert the glucose to lactate, formate, and acetate via pyruvate in the Embden-Meyerhof-Parnas pathway. (See Appendix A in the *Photographic Atlas*.)

As you may remember, lactose is a disaccharide composed of glucose and galactose. *S. thermophilus* does not possess enzymes needed to metabolize galactose, and *L. delbrueckii* preferentially metabolizes glucose. This results in an accumulation of galactose, which adds sweetness to the yogurt. Acetaldehyde is produced directly from pyruvate by *S. thermophilus* and through the conversion of proteolysis products threonine and glycine by *L. delbrueckii*. Some strains of *S. thermophilus* also produce glucose polymers that give the yogurt a viscous consistency.

In this exercise, you will produce yogurt with a simple home recipe using a commercial yogurt, already containing living cultures, as a starter. Read the label to see which microorganisms are included. Hope you enjoy it.

MATERIALS NEEDED FOR THIS EXERCISE

Per Student Group

- Whole, low-fat, or skim milk
- Plain yogurt with active cultures (bring from home or supermarket)
- Medium size saucepan
- Medium size bowl
- Wire whisk
- Hot plate
- Cooking thermometer
- Measuring cup
- Plastic wrap
- Fresh fruit
- Sugar (optional)
- pH meter or pH paper

PROCEDURE

Lab One

1. Obtain all materials and set them up in a clean work area.
2. While stirring, slowly heat 5 cups milk in the saucepan to 185°F. Remove the milk from heat, and let it cool to 110°F.
3. Place 1/4 cup starter yogurt in the bowl. Slowly, about 1/3 to 1/2 cup at a time, stir in cooled milk, mixing after each addition until smooth.
4. Cover the bowl with plastic wrap and puncture several times to allow gases and excess moisture to escape.
5. Label the bowl with your name, the date, and the cultures present in your yogurt starter.
6. Incubate 5–6 hours at 30–35°C. Remove the bowl from the incubator at the correct time and place it in the refrigerator.

Lab Two

1. Remove your yogurt from the refrigerator.
2. Compare flavor, consistency and starter cultures with other groups in the lab. Measure the pH of your yogurt with a pH meter or pH paper. Record your results in the table provided.
3. Serve with fresh fruit and enjoy.

REFERENCES

Downes, Frances Pouch and Keith Ito 2001. Chapter 47 in *Compendium of Methods for the Microbiological Examination of Foods*, 4th Ed. American Public Health Association, Washington, DC.

Ray, Bibek 2001. Chapter 13 in *Fundamental Food Microbiology*, 2nd Ed. CRC Press LLC, Boca Raton, FL.

OBSERVATIONS AND INTERPRETATIONS

OBSERVATIONS AND INTERPRETATIONS			
CULTURE ORGANISMS	FLAVOR	CONSISTENCY	PH

Medical Microbiology

The study and application of microbiological principles and those of medicine are inseparable. So although this is not a medical microbiology manual, it is nonetheless largely devoted to the study of microorganisms and their relationship to human health. Our superficial treatment of medical microbiology in this section is not meant to represent a balanced body of information, but is a list of additional items relating to microbiology and medicine that are important to know.

In this unit you will examine lysozyme—a natural antimicrobial agent produced by the body. You also will perform a test used to detect susceptibility to dental decay.

7-8 LYSOZYME ASSAY

Lysozyme is an enzyme that occurs naturally in egg albumin, and normal body secretions such as tears, saliva, and urine. The enzyme provides limited protection from bacterial infection by breaking certain **peptidoglycan** bonds. Peptidoglycan is made up of cross-linked peptides and the alternating repeating subunits *N*-acetylglucosamine (NAG) and *N*-acetylmuramic acid (NAM). Lysozyme functions by breaking the β-1,4 glycosidic linkages between the NAG and NAM subunits of the glycan polymers.

A lysozyme assay measures the ability of a sample to lyse cells of the substrate organism *Micrococcus lysodeikticus*. Cell lysis occurs as a result of damage caused by the lysozyme and the hypotonic diluent used to dilute the samples.

In this exercise you will measure the lysozyme concentrations in a variety of body fluids. To do this you will first construct a standard curve for comparison. In preparation for the standard curve you will perform a serial dilution of a known concentration of lysozyme, mix the dilutions in equal parts with the bacterial substrate solution, and take a turbidity reading (with a spectrophotometer) of each after 20 minutes.

Greater lysis will occur in high concentrations than in low concentrations of lysozyme; therefore, the range of dilutions and absorbance readings will give you sufficient information to graph the standard curve by plotting light absorbance vs. lysozyme concentration. Dilutions of natural fluid samples can then be prepared, mixed with bacterial substrate solution and read for turbidity after 20 minutes. The light absorbance values of the samples can then be used to interpolate the lysozyme concentrations from the standard curve.

The work for this exercise will be divided among several groups. One group will perform the dilutions of known lysozyme concentrations for construction of the standard curve. Other groups will dilute and test samples of tears, saliva, urine, and/or other body fluids as determined by your instructor.

MATERIALS NEEDED FOR THIS EXERCISE*

- Spectrophotometer
- Timer
- Cuvettes
- Micropipettes (10–100 μL and 100–1000 μL) with sterile tips
- 1 mL and 5 mL pipettes
- Propipettes
- Parafilm
- Lysozyme buffer
- Lysozyme substrate—*Micrococcus lysodeikticus*—in solution measuring 10.0% transmittance
- Lysozyme in the following dilutions:

0.15625	mg/100 mL lysozyme buffer
0.3125	mg/100 mL lysozyme buffer
0.625	mg/100 mL lysozyme buffer
1.25	mg/100 mL lysozyme buffer
2.5	mg/100 mL lysozyme buffer
5.0	mg/100 mL lysozyme buffer

PROCEDURE

Group 1 (Standard Curve)

1. Turn on the spectrophotometer and allow it to warm up for a few minutes. Set the wavelength to 540 nm and, if it is digital, set it to absorbance.
2. Obtain seven cuvettes. Transfer 2 mL of lysozyme substrate (*M. lysodeikticus*) solution to each of six cuvettes; transfer 4 mL straight lysozyme buffer to the seventh.
3. Blank the spectrophotometer with the cuvette containing straight buffer. Continue to check the setting throughout the procedure and blank the machine as needed.

* **Note to instructor.** One liter of buffer is sufficient to make the lysozyme dilutions and substrate solution for a class of 30 to 35 students.

4. Starting with the most dilute (0.15625 mg/100 mL), transfer 2 mL of each lysozyme dilution to the substrate-containing cuvettes respectively.
5. When finished, mark the time. This is T_0.
6. At T_{20} (20 minutes), take turbidity readings of each solution in the same order as in #4. Record your results in the table provided.

Other Groups

1. Turn on the spectrophotometer and allow it to warm up for a few minutes. Set the wavelength to 540 nm and, if it is digital, set it to absorbance.
2. Obtain three cuvettes and all other necessary material for your particular body fluid sample.
3. Transfer 4 mL straight lysozyme buffer to the first cuvette and use it to blank the spectrophotometer. Check this setting frequently throughout the exercise and adjust as necessary.
4. Collect your assigned body fluid according to your teacher's instructions.
5. Add 0.2 mL (200 µL) undiluted sample to a cuvette. Add 1.8 mL lysozyme buffer. Mix. This is a 10^{-1} dilution.
6. Transfer 0.2 mL from the dilution to the second cuvette. Add 1.8 mL lysozyme buffer. Mix. This is 10^{-2}.
7. Add 1.8 mL lysozyme substrate to the first cuvette and 2.0 mL lysozyme substrate to the second cuvette. (This produces 1:1 mixtures of sample and substrate in both tubes.) Mix well. Mark the time. This is T_0.
8. At T_{20} (20 minutes), take turbidity readings of each dilution. Enter your results in the table provided.

■ DATA AND CALCULATIONS

1. Enter data from Group 1.

DATA AND CALCULATIONS STANDARD CURVE

LYSOZYME CONCENTRATION	LIGHT ABSORBANCE T_0	LIGHT ABSORBANCE T_{20}
0.15625 mg/100 mL		
0.3125 mg/100 mL		
0.625 mg/100 mL		
1.25 mg/100 mL		
2.5 mg/100 mL		
5.0 mg/100 mL		

All Groups

2. In the table below, enter class data for all samples.
3. Using the information in the tables, construct a standard curve of light absorbance vs. lysozyme concentration.
4. Using the standard curve, plot the light absorbance values and interpolate the concentrations of each diluted sample.
5. Calculate the original concentrations of all samples, using following formula:

Original Concentration = Concentration of diluted sample × reciprocal of the dilution factor

REFERENCES

DIFCO Laboratories. 1984. Page 515 in *DIFCO Manual*, 10th Ed. DIFCO Laboratories, Detroit, MI.

Sprott, G. Dennis, Susan F. Koval, and Carl A. Schnaitman. 1994. Page 78 in *Methods for General and Molecular Bacteriology*, edited by Philipp Gerhardt, R. G. E. Murray, Willis A. Wood, and Noel R. Krieg, American Society for Microbiology, Washington, DC.

DATA AND CALCULATIONS STANDARD CURVE

BODY FLUID	LIGHT ABSORBANCE T_0		LIGHT ABSORBANCE T_{20}	
	10^{-1}	10^{-2}	10^{-1}	10^{-2}

7-9 SNYDER TEST

Photographic Atlas Reference
Snyder Test Page 96

MATERIALS NEEDED FOR THIS EXERCISE

Per Student Group

- Hot water bath set at 45°C

Per Student

- Small sterile beakers
- Sterile 1 mL pipettes with bulbs
- Two Snyder agar tubes

PROCEDURE

Lab One

1. Collect a small sample of saliva (about 0.5 mL) in the sterile beaker.
2. Aseptically add 0.2 mL of the sample to a molten Snyder agar tube (from the water bath) and roll it between your hands until the saliva is uniformly distributed throughout the agar.
3. Allow the agar to cool to room temperature. Do not slant.
4. Incubate with an uninoculated control at 35°C for up to 72 hours.

Lab Two

1. Examine the tubes at 24-hour intervals for color changes.
2. Record your results in the table provided.

TABLE 7-3 SNYDER TEST RESULTS AND INTERPRETATIONS

RESULT	INTERPRETATION
Yellow at 24 hours	High susceptibility to dental caries
Yellow at 48 hours	Moderate susceptibility to dental caries
Yellow at 72 hours	Slight susceptibility to dental caries
Yellow at >72 hours	Negative

REFERENCES

DIFCO Laboratories. 1984. Page 619 in *DIFCO Manual*, 10th Ed. DIFCO Laboratories, Detroit, MI.

Power, David A. and Peggy J. McCuen. 1988. Page 247 in *Manual of BBL® Products and Laboratory Procedures*, 6th Ed. Becton Dickinson Microbiology Systems, Cockeysville, MD.

■ OBSERVATIONS AND INTERPRETATIONS

Refer to Table 7-3 when recording and interpreting your results below.

OBSERVATIONS AND INTERPRETATIONS

TIME	24 HRS.	48 HRS.	72 HRS.
Color			

Based on the results of this test, I have a ______________________ susceptibility to dental caries.

SECTION EIGHT

8 Microbial Genetics

In this section you will perform three exercises dealing with DNA. First, in Exercise 8-1 you will perform a simple extraction of *E. coli* DNA. In Exercises 8-2 and 8-3, you will examine mutations, that is, alterations of DNA. Exercise 8-2 examines the effects of UV radiation on bacteria. It illustrates some characteristics of a particular **mutagen**—how it causes damage, factors affecting its impact on the cell, and how bacteria are able to repair that damage. Exercise 8-3—the Ames Test—illustrates a simple method of screening substances (*e.g.*, commercial products) to determine mutagenic and potential carcinogenic properties.

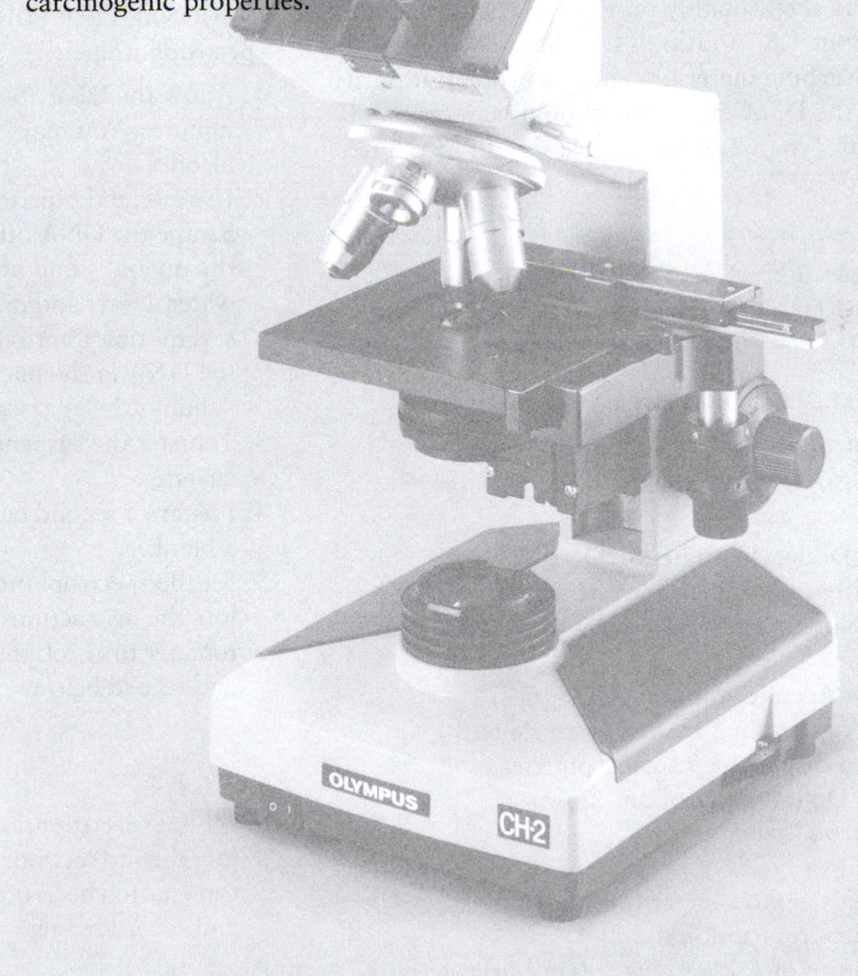

8-1 EXTRACTION OF DNA FROM BACTERIAL CELLS

Extraction of DNA is a starting point for many lab procedures, including DNA sequencing and cloning. The basic process involves three steps: cell lysis, denaturation of protein and other macromolecules, and precipitation of the DNA. The resulting extract should have the consistency of mucus when spooled on a glass rod.

As an optional follow-up to extraction, an ultraviolet spectrophotometer will be used to estimate DNA concentration in the sample by measuring absorbance at 260 nm, the optimum wavelength for absorption by DNA. An absorbance of $A_{260nm} = 1$ corresponds to 50 µg/mL of dsDNA. Purity of the sample can also be determined by reading absorbance at 280 nm and calculating the following ratio:

$$\frac{\text{Absorbance}_{260\text{ nm}}}{\text{Absorbance}_{280\text{ nm}}}$$

If the sample is reasonably pure nucleic acid, then the ratio will be about 1.8. A ratio of less than 1.6 is likely due to other UV-absorbing materials, such as protein. If purity is crucial, then the DNA extraction should be repeated. If the ratio is greater than 2.0, the sample should be diluted and read again.

Photographic Atlas Reference
DNA Extraction Page 105

MATERIALS NEEDED FOR THIS EXERCISE

Per Pair of Students

- Water bath set to 65°C
- Ice bath
- 10% Sodium Dodecyl Sulfate (SDS)
- 20 mg/mL Proteinase K solution (stored in freezer between uses) (Note: meat tenderizer is an inexpensive substitute.)
- 1.0 M Sodium Acetate solution (pH = 5.2)
- 95% Ethanol (stored in a freezer or an ice bath)
- Two calibrated disposable transfer pipettes
- 100–1000 µL Digital pipettor and tips
- Straight glass rod
- Test tube
- Ultraviolet Spectrophotometer (optional)
- Two quartz cuvettes (optional)
- Overnight culture of *Escherichia coli* in Luria-Bertani broth

PROCEDURE

Refer to the Procedural Diagram in Figure 8-1 as you read and perform the following protocol.

1. Obtain the *E. coli* culture. Mix the suspension until a uniform turbidity is seen, then transfer 5 mL to a clean, nonsterile test tube.
2. Add 1 mL of 10% SDS to the *E. coli*.
3. Add 150 µL of 20 mg/mL Proteinase K solution.
4. Gently mix the tube for 5 minutes by rolling it between your hands.
5. Place the tube in a 65°C water bath for 15–30 minutes.
6. After heating, place it in the ice bath until it is at or below room temperature.
7. Add 1 mL 1 M Sodium Acetate solution and mix gently.
8. When cooled, slowly layer 2 mL of cold 95% ethanol onto the surface of the broth. Trickling the ethanol down the side of the tube held at a 45° angle works best.
9. Rotate the glass rod *in one direction* between your hands for about 5 minutes. The DNA will appear as a white, stringy mass at the broth/ethanol interface and will accumulate on the rod as you turn it.

OPTIONAL PROCEDURE (if your lab has an ultraviolet spectrophotometer.)

1. Allow the DNA to dry on the glass rod for a few minutes. You may "pat" it with tissue to remove the alcohol.
2. Use a digital pipettor and obtain 1000 µL of water. Scrape the DNA off the glass into a microtest tube using the tip and simultaneously washing with the 1000 µL of water. Draw and dispense the same 1000 µL of water several times until the glass rod is smooth, then suspend the DNA in the microtest tube until there are no DNA "chunks."
3. Transfer the suspended DNA solution into a quartz cuvette.
4. Prepare a second cuvette containing 1000 µL of water as a blank.
5. Set the spectrophotometer to 260 nm wavelength. Follow the instructions for your particular UV spectrophotometer to check the absorbance of the extracted DNA and record below.

   $ABS_{260} =$

6. Set the spectrophotometer to 280 nm wavelength. Follow the instructions for your particular UV spectrophotometer to check the absorbance of the extracted DNA and record below.

   $ABS_{280} =$

7. Calculate the probable purity of your extracted DNA sample using the following formula:

$$\frac{\text{Absorbance}_{260\text{ nm}}}{\text{Absorbance}_{280\text{ nm}}} =$$

8. If desired, absorbencies at other wavelengths may be taken to produce an absorption spectrum for DNA. Suggested wavelengths are: 200 nm, 220 nm, 240 nm, 300 nm, 320 nm, 340 nm, 360 nm, and 380 nm, in addition to the measurements for 260 and 280 nm taken above. Record these in the table provided.
9. On a separate sheet of graph paper, plot the absorption spectrum (Absorption vs. Wavelength) of the DNA sample.

Observations and Interpretations

Record the absorbance values at each wavelength in the table. Then plot the absorption spectrum (Absorption vs. Wavelength) of the DNA sample on a separate sheet graph paper.

OBSERVATIONS AND INTERPRETATIONS	
Wavelength (nm)	**Absorbance**
200	
220	
240	
260 (from above)	
280 (from above)	
300	
320	
340	

■ Figure 8-1 Procedural Diagram for Bacterial DNA Extraction

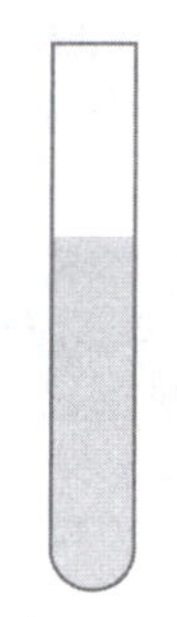

1. 24 hour *E. coli* broth culture.

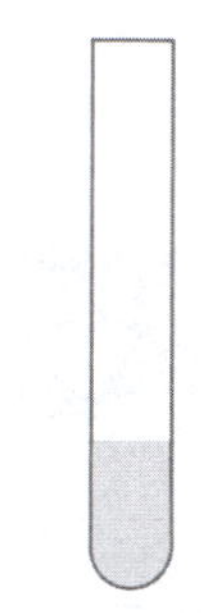

2. Transfer 5 mL *E. coli* to a nonsterile test tube.

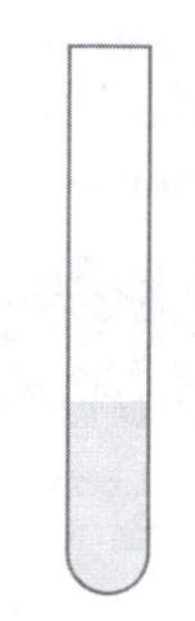

3. Add 1 mL SDS and 150 ∝L Proteinase K, then mix for 5 minutes.

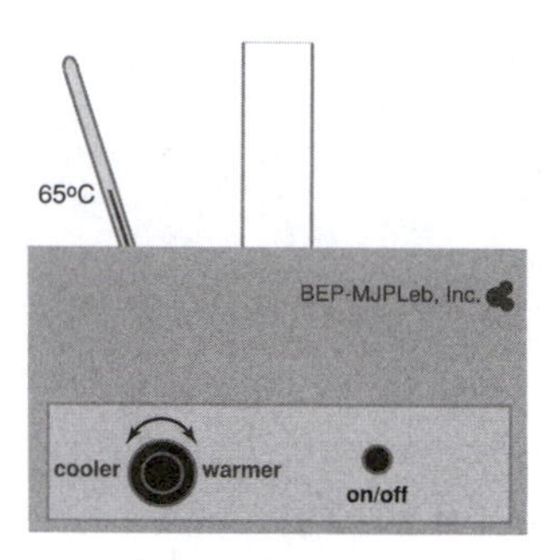

4. Heat for 15-30 minutes at 65ΥC.

5. Put tube in ice bucket until at room temperature. Then add 1 mL Sodim Acetate and mix.

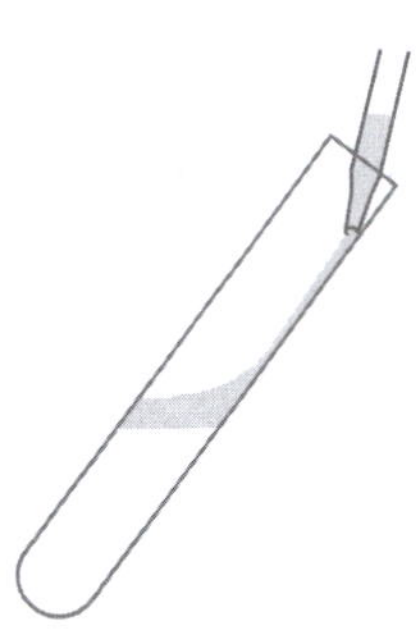

6. Gently layer cold ethanol down the side of the tube.

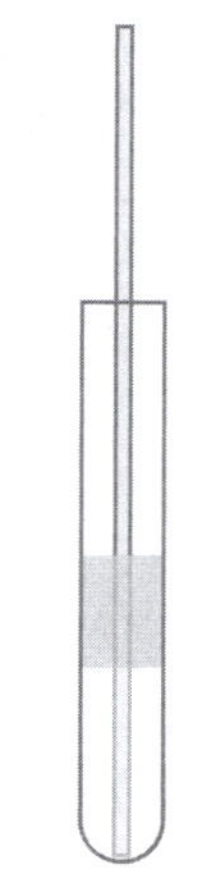

7. Rotate glass rod to collect DNA on it.

REFERENCES

Bost, Rod. 1989. *Down and Dirty DNA Extraction.* Carolina Genes. Research Triangle Park, NC, North Carolina Biotechnology.

Davis, Leonard G., Mark D. Dibner, and James F. Battey. 1986. *Basic Methods in Molecular Biology*. Elsevier Science Publishing Co., Inc. New York, NY.

Kreuzer, Helen and Adrianne Massey 2001. *Recombinant DNA and Biotechnology—A Guide For Teachers*, 2nd Ed., ASM Press, Washington, DC.

8-2 ULTRAVIOLET RADIATION DAMAGE AND REPAIR

This exercise illustrates the effects of ultraviolet radiation (UV) on bacterial cells and allows observation of cellular repair in response to the exposure. It also demonstrates UV's ability to penetrate other matter.

Photographic Atlas Reference
Ultraviolet Radiation: Its Characteristics and Effects on Bacterial Cells Page 109

MATERIALS NEEDED FOR THIS EXERCISE

Per Student Group

- Seven Nutrient Agar plates
- Disinfectant
- Paper or cardboard masks with 1" to 2" cutouts (Figure 8-2)
- UV light (shielded for eye protection)
- Sterile cotton applicators
- 24 hr broth culture of *Serratia marcescens*

PROCEDURE

Lab One

Refer to the Procedural Diagram in Figure 8-3 as you read and perform the following procedure.

1. Using a sterile cotton applicator, streak a nutrient agar plate to form a bacterial lawn over the entire surface by using a tight pattern of streaks. Rotate the plate one-third of a turn and repeat, then rotate it another one-third of a turn and streak one last time. Repeat this process for all remaining plates.
2. Number the plates 1, 2, 3, 4, 5, 6, and 7.
3. Remove the lid from plate 1 and set it on a disinfectant-soaked towel. Place the plate under the UV light and cover it with a mask.
4. Turn the UV light on, but do not look at it. After 30 seconds, turn the UV light off, remove the mask and plate, and replace its lid.
5. Repeat the process for plate 2.
6. Repeat the process for plate 3, but leave the UV light on for 3 minutes.
7. Irradiate plate 4 for 3 minutes, but leave the lid on and cover with the mask.
8. Repeat step 7 for plate 5
9. Do not irradiate plates 6 and 7.
10. Incubate plates 1, 3, 4, and 6 for 24 to 48 hours at room temperature in an inverted position where they can receive natural light (*e.g.*, a windowsill).
11. Wrap plates 2, 5, and 7 in aluminum foil, invert them, and place with the others.

Lab Two

1. Remove the plates and examine the growth patterns. Using the terms "confluent," "individual colonies," and "none," describe the growth of *Serratia marcescens* in both parts of each plate (*i.e.*, where the mask shielded the growth from UV and where it did not).

REFERENCES

Moat, Albert G., John W. Foster, and Michael P. Spector. 2002. Chapter 3 in *Microbial Physiology*, 4th Ed. Wiley-Liss, Inc. New York, NY.

White, David. 2000. Chapter 19 in *The Physiology and Biochemistry of Prokaryotes*. Oxford University Press, Inc. New York, NY.

■ OBSERVATIONS AND INTERPRETATIONS

Record your observations in the table. Examine each region for confluent growth, individual colonies, or no growth. Make a note as to the significance of each type of growth.

OBSERVATIONS AND INTERPRETATIONS

PLATE	MASK/LID	EXPOSURE TIME (MIN)	INCUBATION	GROWTH ON AGAR SURFACE COVERED BY THE MASK	GROWTH ON AGAR SURFACE BENEATH OPENING OF THE MASK
1	Mask, no lid	0.5	Sunlight		
2	Mask, no lid	0.5	Dark		
3	Mask, no lid	3.0	Sunlight		
4	Mask and lid	3.0	Sunlight		
5	Mask and lid	3.0	Dark		
6	Lid only	0	Sunlight		
7	Lid only	0	Dark		

■ Figure 8-2 Cardboard Mask
An example of a cardboard mask placed over a Petri dish (shown as a dotted line). The cutout may be any shape, but should leave the outer 25% of the plate masked.

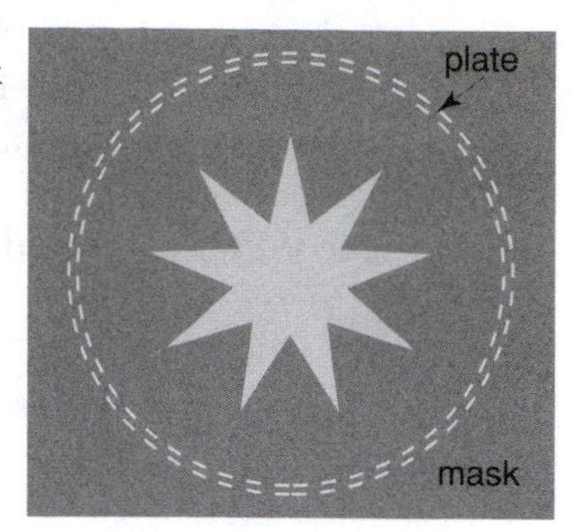

■ Figure 8-3 Procedural Diagram—Ultraviolet Radiation Damage and Repair
Inoculate and expose the plates as directed. Be sure to shield the UV light source adequately. Do not look at the light.

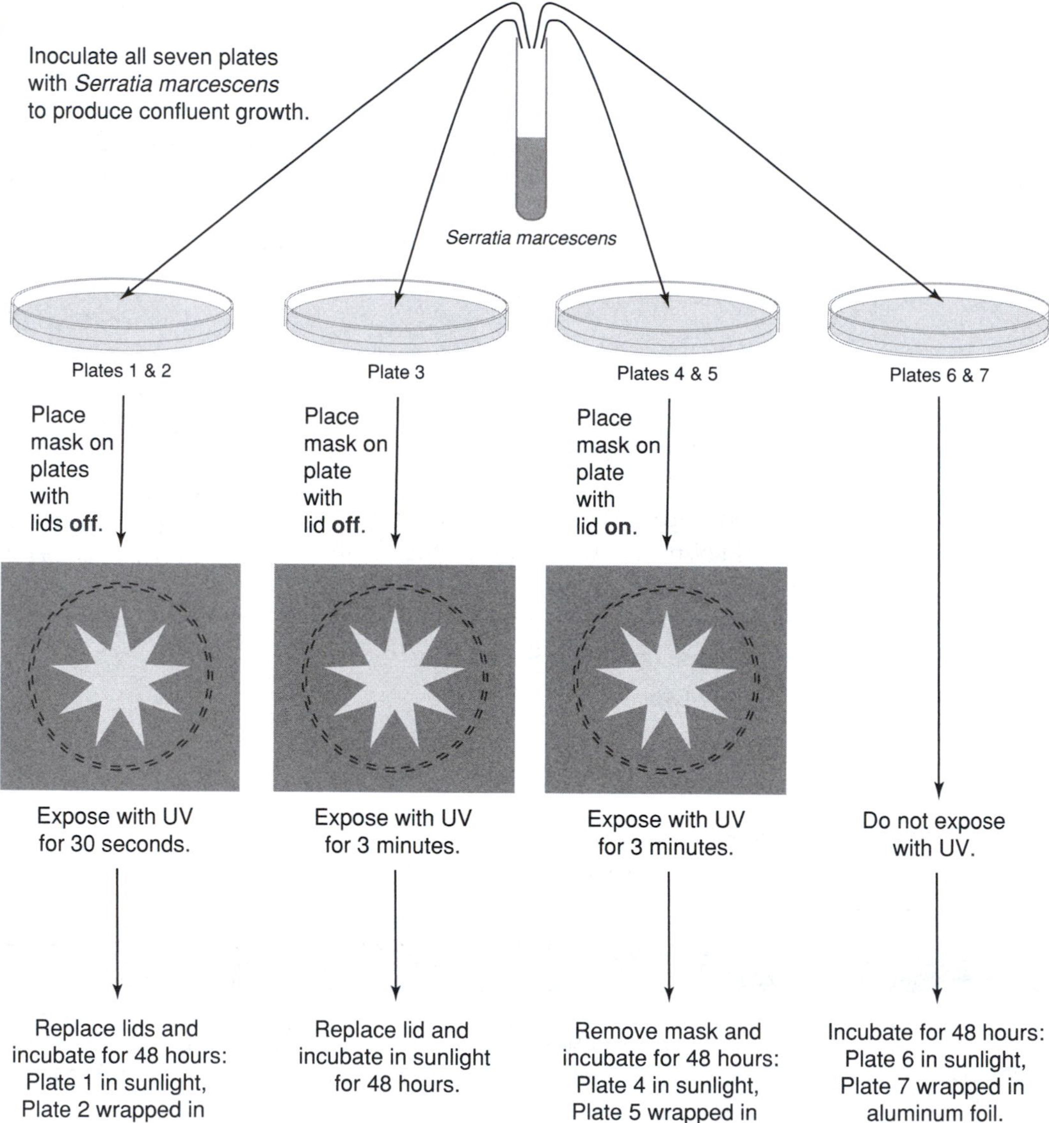

8-3 AMES TEST

Photographic Atlas Reference
Ames Test Page 111

MATERIALS NEEDED FOR THIS EXERCISE

Per Student Group

- Four Minimal Medium (MM) plates
- Four Complete Medium (CM) plates
- Centrifuge
- Two sterile centrifuge tubes
- Small beaker containing alcohol and forceps
- Bottle of 1x Vogel-Bonner salts (To make 1x Vogel-Bonner salts, add 1.0 mL 50x Vogel-Bonner solution to 49 mL water.) (Appendix A)
- Eight sterile filter discs made with a paper punch
- Two sterile Petri dishes (for soaking filter paper discs)
- Two sterile transfer pipettes
- 100 µL–1000 µL digital pipettors and tips
- Container for disposal of supernatant (to be autoclaved)
- DMSO
- Test substance (any substance that has possible mutagenic properties and does not contain histidine or protein)
- Broth culture* of *Salmonella typhimurium TA 1535*
- Broth culture* of *Salmonella typhimurium TA 1538*

* The cultures used for this exercise must be prepared as follows:
1. 24 hours before the test, inoculate two 10.0 mL broth tubes with TA 1535 and TA 1538. Incubate at 35°C together with two sterile 90.0 mL broths (in a 100mL diluent bottle).
2. 5½ hours before the test, pour the TA 1538 culture into one of the sterile 90.0 mL broths and return it to the incubator until time for the exercise.
3. 4 hours before the test, pour the TA 1535 culture into the other 90.0 mL sterile broth and return it to the incubator until time for the exercise.

PROCEDURE

Day One

Follow the Procedural Diagram in Figure 8-4.

1. Soak four filter paper discs in DMSO and four filter paper discs in the test substance.
2. Pipette 10.0 mL TA 1535 into a sterile centrifuge tube. Do the same with TA 1538, then label the tubes.
3. Centrifuge the tubes on high speed for 10 minutes. Be sure the centrifuge is balanced.
4. Being careful not to disturb the cell pellet at the bottom, decant the supernatant from each centrifuge tube using a transfer pipette.
5. Resuspend the cell pellets by adding l.0 mL sterile 1x Vogel-Bonner salts to each tube and mixing well.
6. Using spread plate technique (see Exercise 1-4) inoculate two MM plates and two CM plates with 100 µL of the resuspended TA 1535. Do the same with TA 1538.
7. Flame the forceps by passing them through the Bunsen burner flame and allowing the alcohol to burn off.
8. Using the flamed forceps, place the DMSO and test discs in the centers of the plates. Gently tap the disks down with the forceps to prevent them falling off when the plates are inverted.
9. Incubate the plates aerobically at 35°C for 48 hours.

Day Two

1. Measure and compare the zones of inhibition (measured in millimeters) on the CM plates.
2. Count the colonies on the MM plates and compare. Count only the large colonies, not the "hazy," background growth. (These are the colonies produced by auxotrophs that did not back-mutate to prototrophs and only grew until the histidine in the minimal medium was exhausted.)
3. Record your observations for the Ames test plates below.

■ OBSERVATIONS AND INTERPRETATIONS

Examine the CM plates and measure the zones of inhibition around the disks. Examine the MM plates and count the colonies. Then compare the DMSO plates with the Test Substance plates.

OBSERVATIONS AND INTERPRETATIONS

	ZONE DIAMETER		COLONIES COUNTED	
	CM WITH TA 1535	CM WITH TA 1538	MM WITH TA 1535	MM WITH TA 1538
DMSO				
Test Substance				

■ **Figure 8-4 Procedural Diagram**
Be sure to dispose of all pipettes, broth, and tubes properly as you perform the experiment. After 48 hours of incubation, use a metric ruler to measure the diameter of the cleared zone on all CM plates. Use a colony counter to count the back-mutant colonies on all the MM plates. (**Note:** These will be fairly large. Do not count the tiny, hazy growth over the plate's surface.) Mark each colony with a toothpick (to avoid counting it more than once) as you keep track of the number using a hand tally counter.

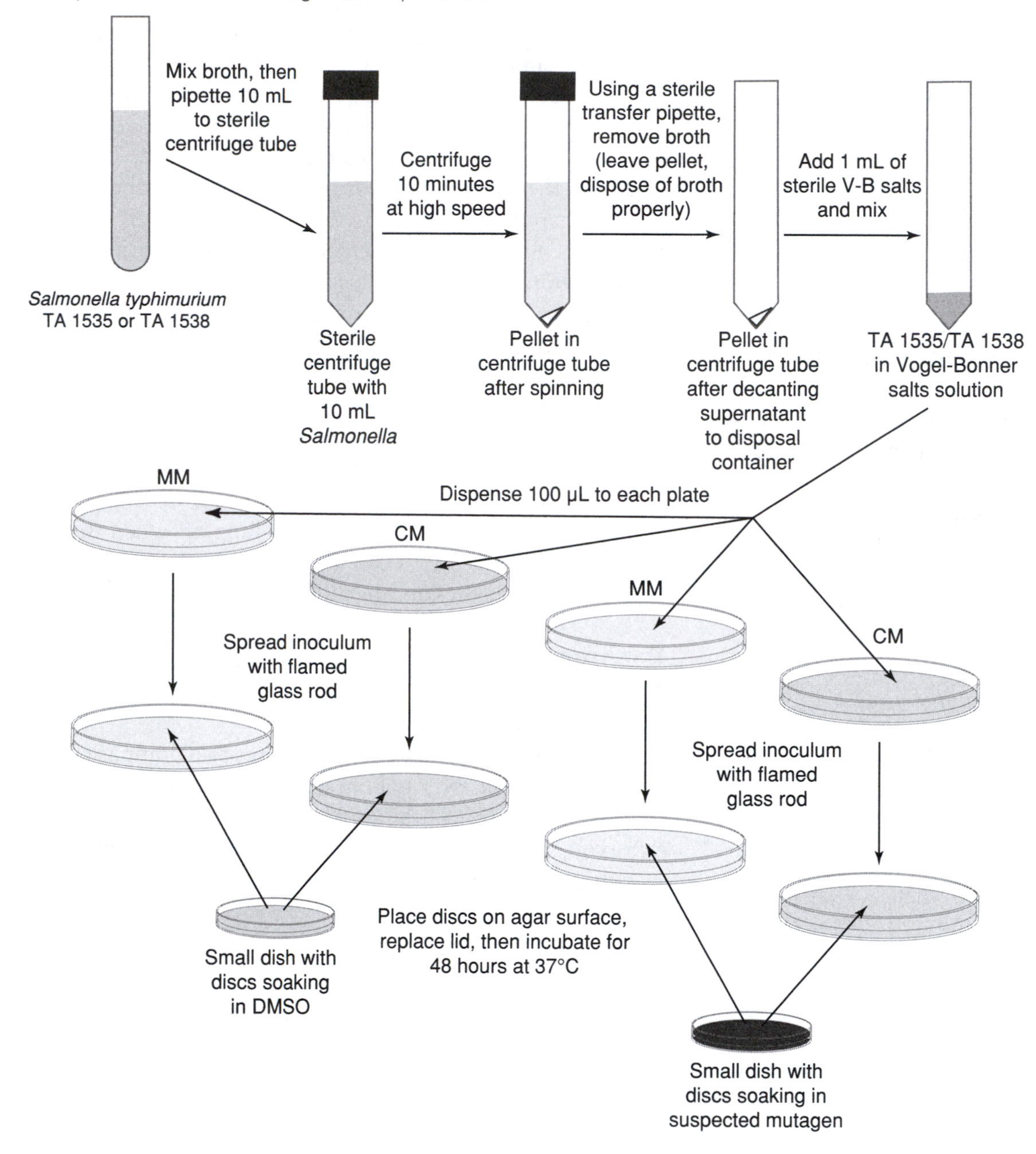

REFERENCES

Eisenstadt, Bruce C. Carlton, and Barbara J. Brown. 1994. Page 311 in *Methods for General and Molecular Bacteriology*, edited by Philipp Gerhardt, R. G. E. Murray, Willis A. Wood, and Noel R. Krieg, American Society for Microbiology, Washington, DC.

Maron, D. M. and B. N. Ames. 1983. *Mutation Research,* 113: 173–215.

SECTION NINE

9 Hematology and Serology

This section deals with blood cells and other aspects of the body's defenses. In Exercise 9-1 you will have the opportunity to perform a differential blood cell count. Exercises 9-2 through 9-5 allow you to perform serological tests used to detect the presence of specific antigens or antibodies in a sample.

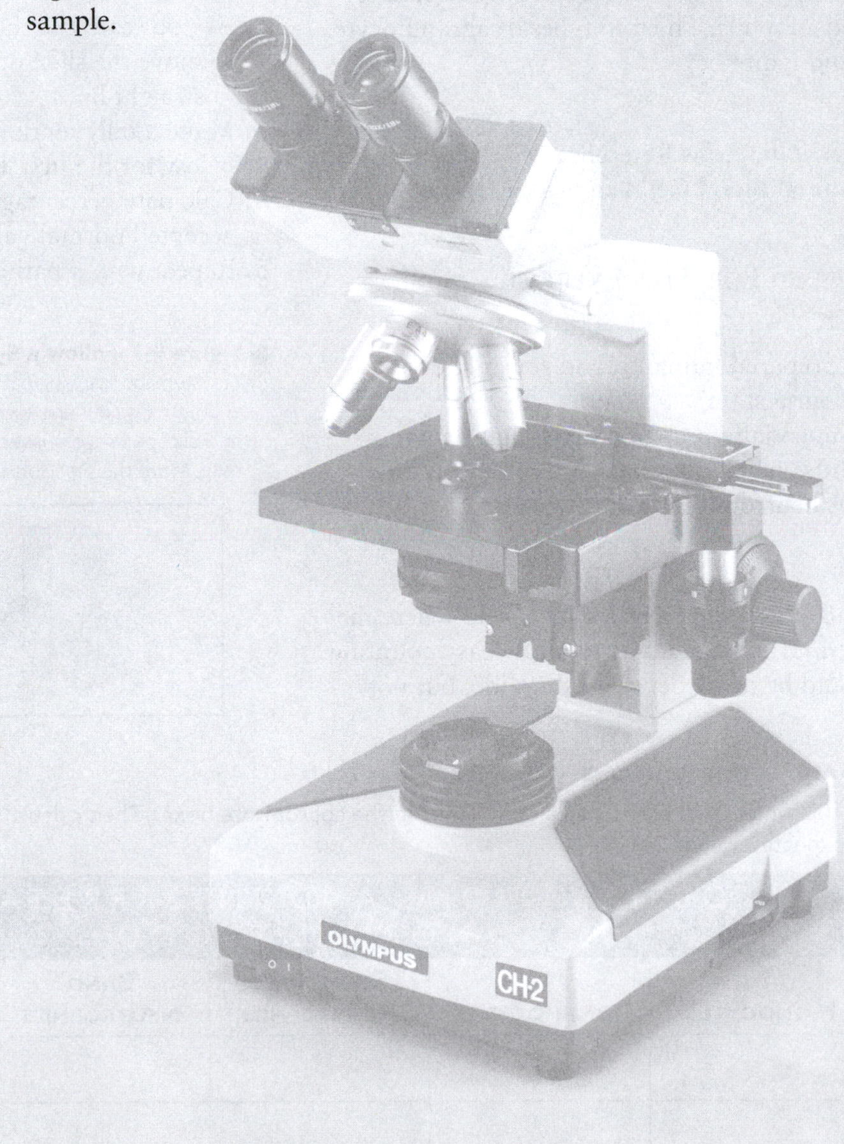

Hematology and Immunology

The first exercise in this section deals with the blood. In Exercise 9-1, a differential blood cell count will be done. Clinical laboratories do these as a means of diagnosing (or eliminating from consideration) certain pathological conditions. While differential counts are automated now, it is good training to perform one "the old-fashioned way" using a blood smear and a microscope to get an idea of the principle behind the technique.

9-1 DIFFERENTIAL BLOOD CELL COUNT

Leukocytes (white blood cells or WBCs) are divided into two groups: **granulocytes** (which have prominent cytoplasmic granules) and **agranulocytes** (which lack these granules). There are three basic types of granulocytes: neutrophils, basophils, and eosinophils. The two types of agranulocytes are monocytes and lymphocytes.

Photographic Atlas Reference
Differential Blood Cell Count Page 99

MATERIALS NEEDED FOR THIS EXERCISE

Per Student

- Commercially prepared human blood smear slides (Wright's or Giemsa stain)
- Optional: Commercially prepared abnormal human blood smear slides (*e.g.,* infectious mononucleosis, eosinophilia, or neutrophilia)

PROCEDURE

1. Obtain a blood smear slide and locate a field where the cells are spaced far enough apart to allow easy counting. (The cells should be fairly dense on the slide, but not overlapping.)
2. Using the oil immersion lens, scan the slide using the pattern shown in Figure 9-1. Be careful not to overlap fields when scanning the specimen. Choose a "landmark" blood cell at the right side of the field and move the slide horizontally until that cell disappears off the left side. Also, avoid diagonal movement of the slide. As you scan, use the mechanical stage knobs separately to move the slide up and back or to the right and left in straight lines.
3. Make a tally mark in the appropriate box in the table below for the first 100 leukocytes you see.
4. Calculate percentages and compare your results with the accepted normal values.
5. Repeat with a pathological blood smear (if available).

Figure 9-1 Follow a Systematic Path
A systematic scanning path is used to avoid wandering around the slide and perhaps counting some cells more than once. Remember that a microscope image is inverted. If you want the image to move left, you must move the slide to the right.

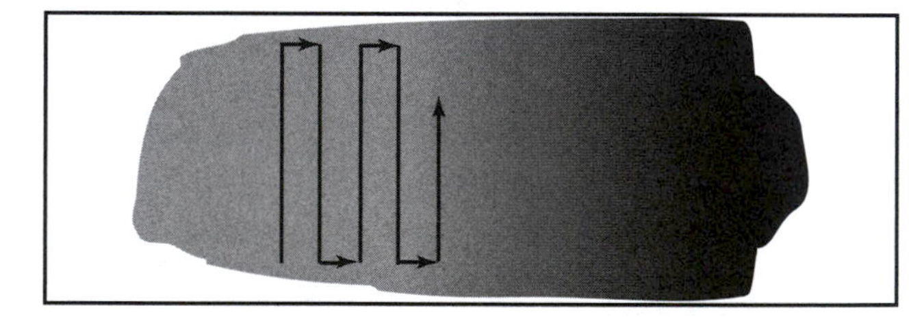

OBSERVATIONS AND INTERPRETATIONS

As you count the 100 white blood cells, make tally marks in the appropriate boxes. Then calculate the percentages of each type and compare them to the expected values.

OBSERVATIONS AND INTERPRETATIONS NORMAL BLOOD

	MONOCYTES	LYMPHOCYTES	SEGMENTED NEUTROPHILS	BAND NEUTROPHILS	EOSINOPHILS	BASOPHILS
Number						
Percentage						
Expected Percentage	3–7%	25–33%	55–65% (All neutrophils)	—	1–3%	0.5–1%

OBSERVATIONS AND INTERPRETATIONS
ABNORMAL BLOOD

CONDITION (________)	MONOCYTES	LYMPHOCYTES	SEGMENTED NEUTROPHILS	BAND NEUTROPHILS	EOSINOPHILS	BASOPHILS
Number						
Percentage						
Expected Percentage	3–7%	25–33%	55–65% (All neutrophils)	—	1–3%	0.5–1%

REFERENCES

Brown, Barbara A. 1993. *Hematology—Principles and Procedures,* 6th Ed. Lea and Febiger, Philadelphia, PA.

Diggs, L. W., Dorothy Sturm, and Ann Bell. 1978. *The Morphology of Human Blood Cells*, 4th Ed. Abbott Laboratories, North Chicago, IL.

Junqueira, L. Carlos, and Jose Carneiro. 2003. *Basic Histology, Text and Atlas*, 10th Ed. Lange Medical Books, McGraw Hill, New York, NY.

Leboffe, Michael J. 2003. *A Photographic Atlas of Histology*. Morton Publishing Company, Englewood, CO.

Simple Serological Reactions

Antigen-antibody reactions are very specific, and occur *in vitro* as well as *in vivo*. Serology is the discipline that exploits this specificity as an *in vitro* diagnostic tool. Two simple serological reactions—agglutination and precipitation—are used in the following exercises because they result in the formation of complexes that can be viewed with the naked eye and without sophisticated equipment.

The precipitin ring test (Exercise 9-2) and the radial immunodiffusion test (Exercise 9-3) illustrate precipitation. Both may be used to identify antigens or antibodies in a sample; the latter is also used to compare antigens in more than one sample. The slide agglutination test (Exercise 9-4) can be an important diagnostic (and very specific) tool for the identification of organisms. It is especially useful for serotyping large genera such as *Salmonella*. Hemagglutination, which detects specific antigens on RBCs, is the standard test for determining blood type (Exercise 9-5). Other hemagglutination tests are used for diagnosis of infections.

9-2 PRECIPITIN RING TEST

Photographic Atlas Reference
Precipitin Ring Test on Page 113

Materials Needed for this Exercise

Per Student Group

- Two clean 6 × 50 mm Durham tubes
- Equine serum (containing equine albumin)
- Equine albumin antiserum (containing equine albumin antibodies)
- 0.9% saline solution
- Pasteur pipettes

Procedure

1. Carefully add equine antiserum to both Durham tubes. Fill from the bottom of the tube until it is about 1/3 full.
2. Mark one tube "A". Add the equine serum in such a way that a sharp and distinct second layer is formed without any mixing of the two solutions. It is critical not to allow any mixing. Success can usually be achieved by allowing the serum to slowly trickle down the inside of the glass (Figure 9-2).
3. Mark the second tube "B". Add the 0.9% saline the same way you added equine serum to tube A.
4. Incubate both at 35°C undisturbed for one hour.
5. Observe the tubes for the characteristic ring formed at the interface of the two solutions (zone of optimum proportions). If after one hour there is no ring in either tube, place them in the refrigerator for 12 to 24 hours and then recheck.

Observations and Interpretations

Sketch and label the two tubes. Indicate any line(s) of precipitation.

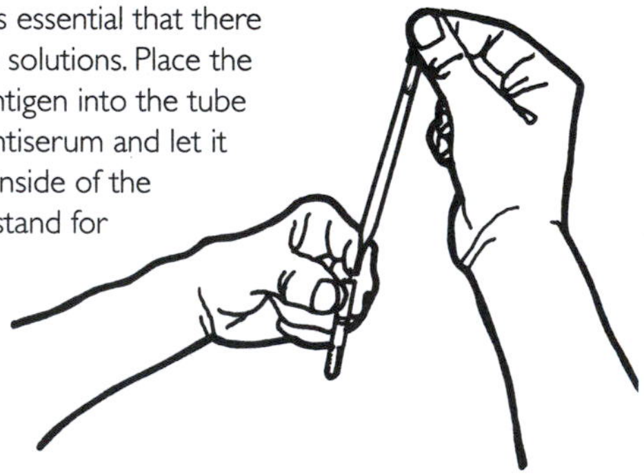

Figure 9-2 Layering the Antigen On Top of the Antibody
When adding the antigen layer to the tube containing antiserum, it is essential that there be no mixing of the two solutions. Place the pipette containing the antigen into the tube about 5 mm from the antiserum and let it slowly trickle down the inside of the glass. Allow the tube to stand for one hour undisturbed.

Reference

Lam, Joseph S. and Lucy M. Mutharia. 1994. Page 120 in *Methods for General and Molecular Bacteriology*, edited by Philipp Gerhardt, R. G. E. Murray, Willis A. Wood, and Noel R. Krieg, American Society for Microbiology, Washington, DC.

9-3 GEL IMMUNODIFFUSION

Photographic Atlas Reference
Precipitation Reactions Page 113

MATERIALS NEEDED FOR THIS EXERCISE

Per Student Group

- One Saline Agar plate
- 3 mm punch (a glass dropper with a 3 mm diameter tip will work)
- Template for cutting wells. Use the template in Observations and Interpretation.
- Bovine antiserum (containing bovine albumin antibodies)
- Equine antiserum (containing equine albumin antibodies)
- 10% bovine serum (Prepared by adding 9 drops physiological saline to 1 drop 100% serum)
- 10% equine serum (Prepared by adding 9 drops physiological saline to 1 drop 100% serum)
- 0.9% saline solution
- Disposable micropipettes
- Two small test tubes

PROCEDURE

Lab One

1. Center the saline agar plate over the template in Observations and Interpretation.
2. Using the punch or glass dropper, cut the 7 wells in the agar as shown in Figure 9-3. If the small agar disks don't come out with the dropper, use suction with the dropper and bulb to dislodge and remove them. Avoid lifting the agar when removing the punched-out disks.
3. Number the peripheral wells 1 through 6 as shown in Observations and Interpretations.
4. In a small test tube or mixing cup prepare a 1:1 mixture of equine albumin antiserum and bovine albumin antiserum. In a second test tube, make a 1:1 mixture of equine and bovine serum.
5. Using a *different* micropipette for each transfer, carefully fill the wells. Place the tip of the pipette in the bottom of the well. Fill slowly to prevent creating air bubbles and to minimize spilling serum over the sides.
   a. Fill the center well with the equine/bovine antiserum mixture.
   b. Fill wells 1 and 2 with 10% equine serum.
   c. Fill wells 3 and 4 with 10% bovine serum.
   d. Fill well 5 with 0.9% saline.
   e. Fill well 6 with the equine/bovine serum mixture.
6. Cover the plate and allow it to sit undisturbed for 30 minutes.
7. Incubate at room temperature for up to 72 hours or until precipitation lines appear.

■ Figure 9-3 Boring Wells In the Saline Agar Plate
When cutting wells in the agar, press straight down; do not twist. Twisting the cutter or dropper may create fissures in the agar, which could disrupt the diffusion of the solutions.

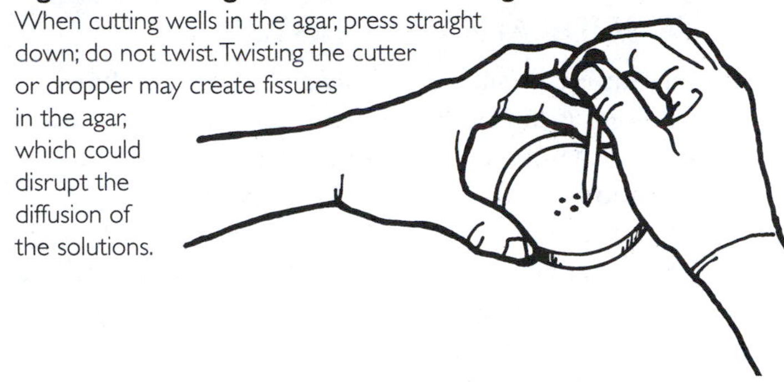

Lab Two

1. Examine the plate for precipitation lines (Figure 9-4) and draw them below. For each pair of wells, interpret the precipitation patterns.

■ Figure 9-4 Precipitation Patterns
The diagram shows three possible precipitation patterns formed between the antibody in the center well and antigens in wells #1 and #2. The pattern on the left demonstrates identity between the antigens. This indicates that the antigens are identical. The pattern in the middle demonstrates partial identity, which means that the antigens are related but not identical (they share some epitopes). The pattern on the right shows nonidentity; the antigens are not related.

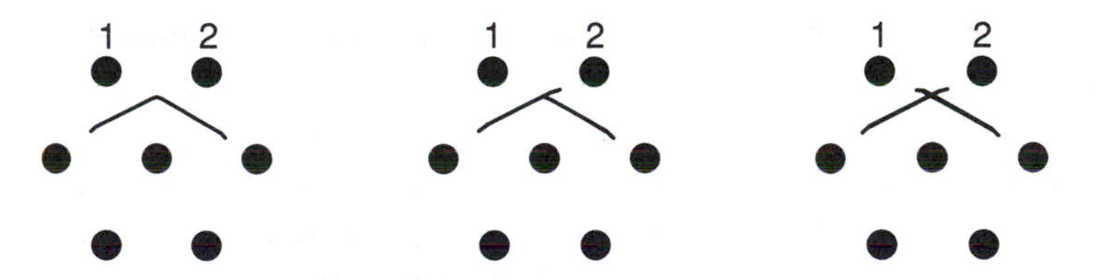

■ OBSERVATIONS AND INTERPRETATIONS

Draw the pattern of precipitation lines between the wells. Interpret each combination as "identity," "partial identity," or "nonidentity."

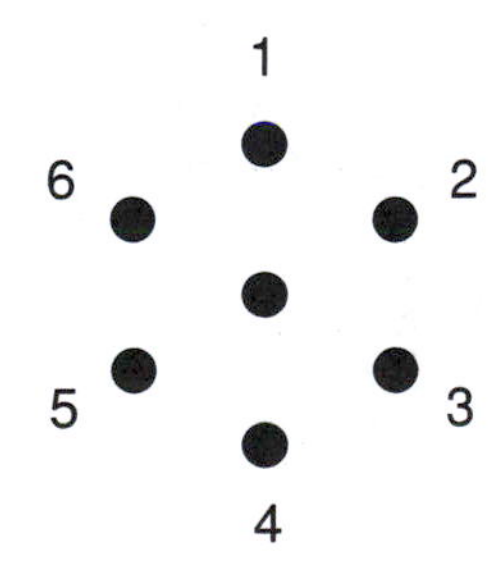

REFERENCES

Lam, Joseph S. and Lucy M. Mutharia. 1994. Page 120 in *Methods for General and Molecular Bacteriology*, edited by Philipp Gerhardt, R. G. E. Murray, Willis A. Wood, and Noel R. Krieg, American Society for Microbiology, Washington, DC.

Ouchterlony, O. 1968. Page 20 in *Handbook of Immunodiffusion and Immunoelectrophoresis*. Ann Arbor Science Publishers, Ann Arbor, MI.

9-4 SLIDE AGGLUTINATION

Particulate antigens (such as whole cells) will combine with homologous antibodies to form visible clumps called **agglutinates. Agglutination** thus serves as evidence of antigen-antibody reaction and is considered a positive result. Agglutination reactions are very sensitive and may be used to detect either the presence of antigen or antibody in a sample.

Photographic Atlas Reference
Agglutination Reactions Page 114

MATERIALS NEEDED FOR THIS EXERCISE

Per Student Group

- *Salmonella* H antigen
- *Salmonella* O antigen
- *Salmonella* anti-H antiserum
- One clean microscope slide
- Toothpicks
- Marking pen

PROCEDURE

1. Using a marking pen, draw two circles approximately the size of a dime on a microscope slide. Label one "O" and the other "H" (Figure 9-5).
2. Place a drop of *Salmonella* anti-H antiserum in each circle.
3. Place a drop of *Salmonella* O antigen in the "O" circle and a drop of *Salmonella* H antigen in the "H" circle. Be careful not to touch the dropper to the antiserum already on the slide.
4. Using a *different* toothpick for each circle, mix until each of the antigens is completely emulsified with the antiserum. Do not over mix. Discard the toothpicks in a biohazard container.
5. Allow the slide to sit for a few minutes and observe for agglutination. Record your results in Observations and Interpretations.

■ Figure 9-5 Prepare the Slide
With your marking pen, draw two dime-sized circles on the slide. Label one circle "O" and the other "H". The circles will be where you check for the presence of Salmonella O and H antigens, respectively.

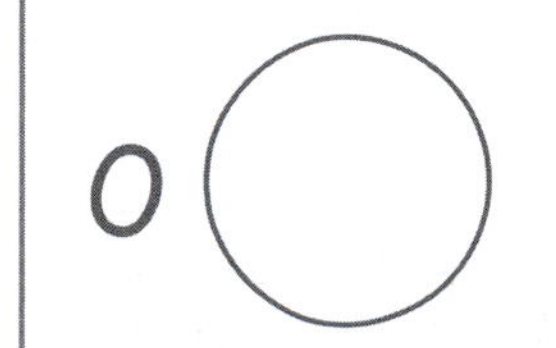

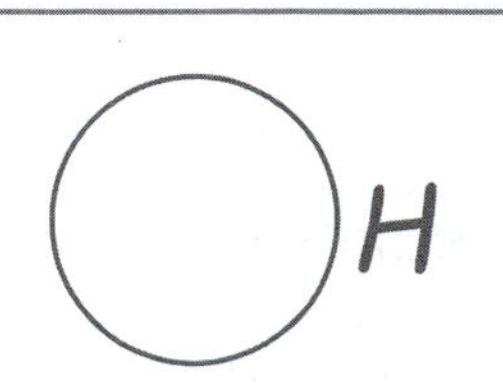

ALTERNATE TEST PROCEDURE (UNKNOWN ANTIGEN)

Your instructor will cover the labels of the two *Salmonella* antigen bottles. You will use this procedure to identify which one contains the *Salmonella* H antigen.

1. Using a marking pen, draw two circles approximately the size of a dime on a microscope slide (Figure 9-5).
2. Place a drop of one *Salmonella* unknown antigen in one circle and a drop of the other *Salmonella* unknown antigen in the other circle (Figure 9-5).
3. Place a drop of *Salmonella* anti-H antiserum in each circle. Be careful not to touch the dropper to the antigen solutions already on the slide.
4. Using a *different* toothpick for each circle, mix until each of the antigens is completely emulsified with the antiserum. Discard the toothpicks in a biohazard container.
5. Allow the slide to sit for a few minutes and observe for agglutination.

■ OBSERVATIONS AND INTERPRETATIONS

Sketch your results in one of the following diagrams.

Known Antigens

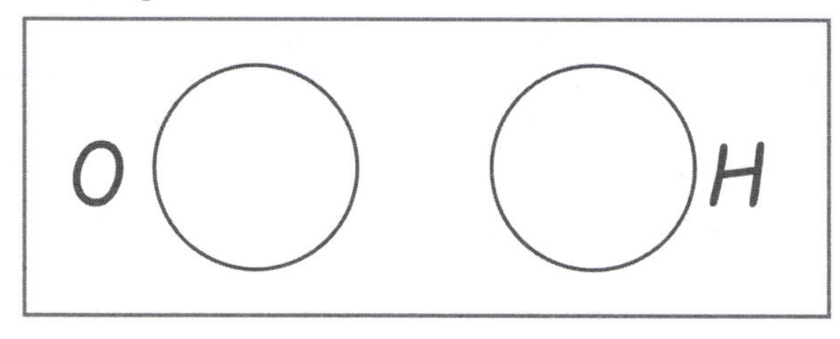

Unknown Antigens
Determine which sample contained Samonella H antigen and label appropriately.

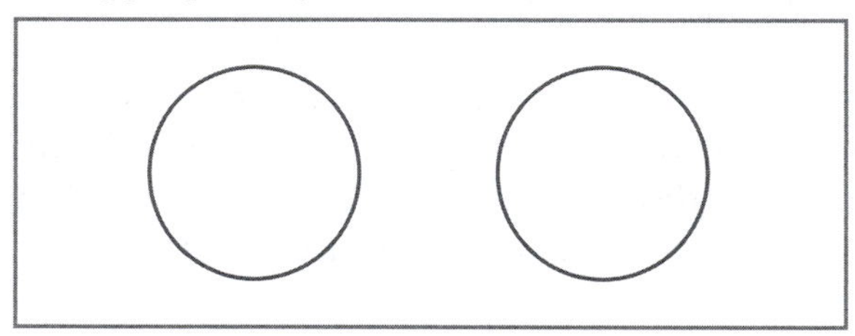

REFERENCES

Bopp, Cheryl A., Frances W. Brenner, Joy G. Wells, and Nancy A. Strockbine. 1999. Pages 467–471 in *Manual of Clinical Microbiology*, 7th Ed., edited by Patrick R. Murray, Ellen Jo Baron, Michael A. Pfaller, Fred C. Tenover, and Robert H. Yolken. American Society for Microbiology, Washington, DC.

Collins, C. H., Patricia M. Lyne, J. M. Grange. 1995. Page 118 in *Collins and Lyne's Microbiological Methods*, 7th Ed. Butterworth-Heinemann, UK.

Constantine, Niel T. and Dolores P. Lana. 2003. Pages 222–223 in *Manual of Clinical Microbiology*, 8th Ed., edited by Patrick R. Murray, Ellen Jo Baron, James H. Jorgensen, Michael A. Pfaller, and Robert H. Yolken. American Society for Microbiology, Washington, DC.

Forbes, Betty A., Daniel F. Sahm, Alice S. Weissfeld. 2002. Pages 206–207 in *Bailey & Scott's Diagnostic Microbiology*, 11th Ed. Mosby, Inc., St. Louis, MO.

Lam, Joseph S. and Lucy M. Mutharia. 1994. Page 120 in *Methods for General and Molecular Bacteriology*, edited by Philipp Gerhardt, R. G. E. Murray, Willis A. Wood, and Noel R. Krieg, American Society for Microbiology, Washington, DC.

9-5 BLOOD TYPING

Hemagglutination is a general term applied to any agglutination test in which clumping of red blood cells indicates a positive reaction. Blood tests as well as a number of indirect diagnostic serological tests are hemagglutinations.

Photographic Atlas Reference
Agglutination Reactions Page 115

MATERIALS NEEDED FOR THIS EXERCISE

Per Student Group

- Blood typing anti-A antiserum
- Blood typing anti-B antiserum
- Blood typing anti-Rh (anti-D) antiserum
- Two microscope slides
- Toothpicks
- Marking pen
- Sterile lancets
- Alcohol wipes
- Small adhesive bandages
- Sharps container
- Disposable gloves

PROCEDURE

1. Draw two circles with your marking pen on one microscope slide. Label one circle "A" and the other "B" (Figure 9-6a).
2. Draw a single circle with your marking pen in the center of a second microscope slide. Label it "Rh" (Figure 9-6b).

■ Figure 9-6 Prepare the slides
(a) Draw two circles on a slide and label them "A" and "B" for the type of antiserum each will receive. (b) Draw one circle on a second slide and label it "Rh."

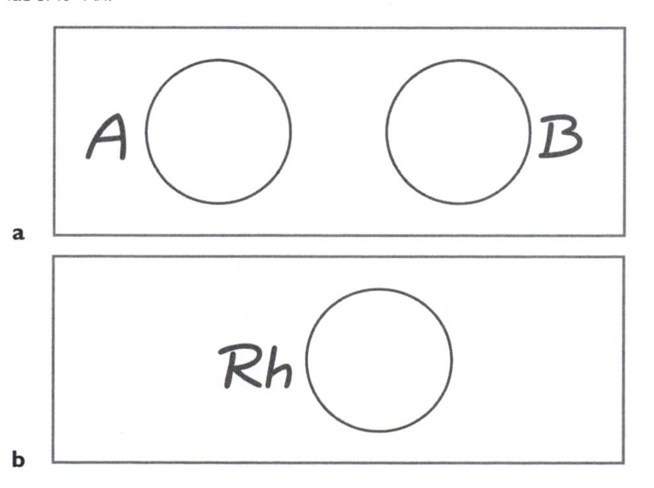

3. Place a drop of anti-A antiserum in the "A" circle.
4. Place a drop of anti-B antiserum in the "B" circle.
5. On the second microscope slide, place a drop of anti-Rh antiserum.
6. Clean the tip of your finger with an alcohol wipe. Let the alcohol dry.
7. Open a lancet package and remove the lancet, being careful not to touch the tip before you use it.
8. Prick the end of your finger and immediately place a drop of blood beside each drop of antiserum. Do not touch the antisera with your finger. It's OK to have someone else prick your finger, but make sure he or she wears protective gloves.
9. Discard the lancet in the sharps container.
10. Put an adhesive bandage on your wound.
11. Using a circular motion, mix each set of drops with a toothpick. Be sure to use a *different toothpick* for each antiserum.
12. Gently rock the slides back and forth for a few minutes or until agglutination occurs.
13. After the agglutination reaction is complete, record the results in the table provided. Compare your results with the possible results in Table 9-1 to determine your blood type.
14. Collect class data and record these in the table provided.

■ OBSERVATIONS AND INTERPRETATIONS

Record your results in the table below. Then collect class data, calculate percentages and compare to the national values.

OBSERVATIONS AND INTERPRETATIONS

ANTISERUM	AGGLUTINATION +/–
Anti-A	
Anti-B	
Anti-Rh	

My blood type is: ______________

REFERENCE

American Association of Blood Banks Website. 2000. http://www.aabb.org/All_About_Blood/FAQs/aabb_faqs.htm#Facts

TABLE 9-1 INTERPRETING BLOOD TYPES

Anti-A Antiserum	Anti-B Antiserum	Anti-Rh Antiserum	Interpretation	Symbol
Agglutination	No Agglutination	Agglutination	A antigen present Rh antigen present	A+
		No Agglutination	A antigen present Rh antigen absent	A–
No Agglutination	Agglutination	Agglutination	B antigen present Rh antigen present	B+
		No Agglutination	B antigen present Rh antigen absent	B–
Agglutination	Agglutination	Agglutination	A and B antigens present Rh antigen present	AB+
		No Agglutination	A and B antigens present Rh antigen absent	AB–
No Agglutination	No Agglutination	Agglutination	A and B antigens absent Rh antigen present	O+
		No Agglutination	A and B antigens absent Rh antigen absent	O–

OBSERVATIONS AND INTERPRETATIONS

Blood Type	Percentage in U.S. Population*	Number in Class	Percentage in Class	Deviation from National Values
O+	38			
O–	7			
A+	34			
A–	6			
B+	9			
B–	2			
AB+	3			
AB–	1			

*Source: American Association of Blood Banks;

SECTION TEN

10 Eukaryotic Microbes: Fungi, Protozoans, and Helminths

College microbiology courses are usually dominated by bacteriology, but the discipline also includes eukaryotic microorganisms. You have had the opportunity to look at some of these organisms in Exercise 3-3. In this section, you will continue to examine some representative microscopic fungi (Exercise 10-1) and protozoans (Exercise 10-2) that are of medical or commercial importance. In Exercise 10-3, you will examine parasitic helminths (worms). While these are not microorganisms, clinical microbiologists often encounter them or their eggs as they examine various patient samples. For this reason, helminth parasites have entered the domain of microbiology.

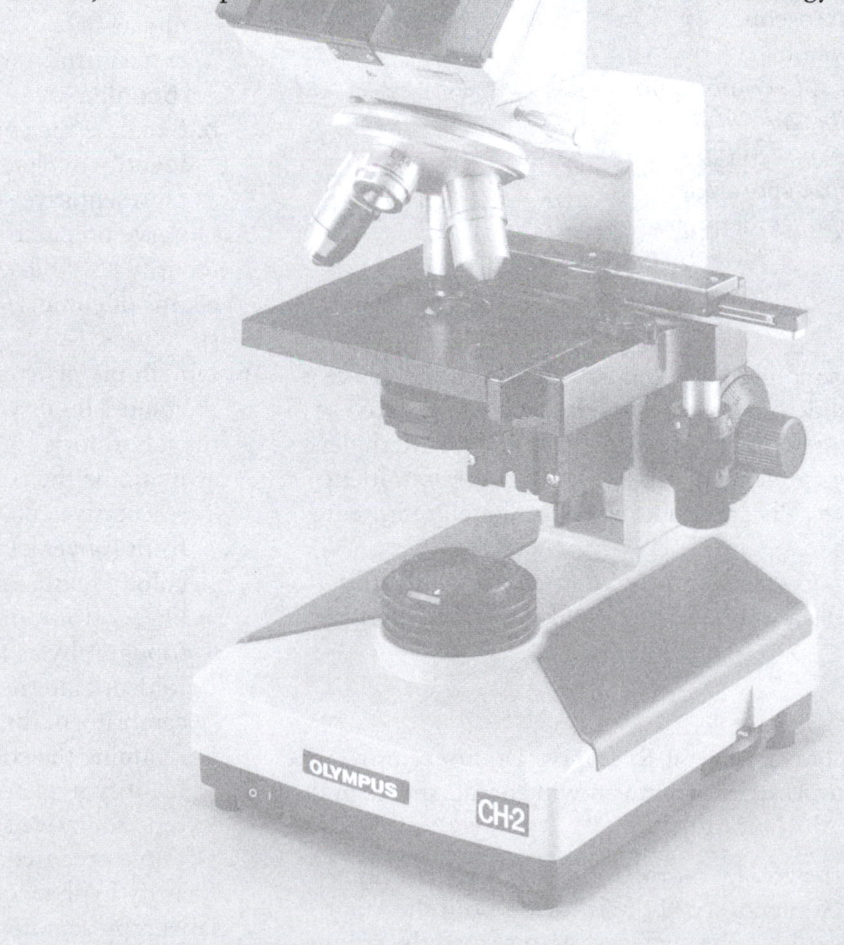

10-1 FUNGI

Members of the Kingdom Fungi are eukaryotic absorptive heterotrophs. While formally divided into the Divisions Ascomycetes, Zygomycetes, and Deuteromycetes, we take the simpler approach of dividing them into the unicellular **yeasts** and filamentous **molds**.

Photographic Atlas Reference
Survey of Fungi Page 159

Material Needed For This Exercise (per group)

Per Student Group

- Agar slant of *Saccharomyces cerevisiae*
- Potato Dextrose Agar or Sabouraud Dextrose Agar plate culture of *Aspergillus spp.* (with lid taped on)
- Potato Dextrose Agar or Sabouraud Dextrose Agar plate culture of *Penicillium spp.* (with lid taped on)
- Potato Dextrose Agar or Sabouraud Dextrose Agar plate culture of *Rhizopus spp.* (with lid taped on)
- Gram's iodine stain or Methylene Blue
- Dissecting microscope
- Prepared slides of:
  - *Aspergillus* spp. conidiophore
  - *Candida albicans*
  - *Penicillium spp.* conidiophore
  - *Rhizopus spp.* sporangia
  - *Rhizopus spp.* gametangia

Procedure

Yeasts

1. Using an inoculating loop and aseptic technique, make a wet mount slide of *Saccharomyces cerevisiae* as illustrated in Figure 3-18 and stain with iodine or Methylene Blue. Observe under high dry and oil immersion. Identify vegetative cells and budding cells. Sketch representative cells.
2. Observe prepared slides of *Candida albicans*. Identify vegetative cells and budding cells. Sketch representative cells.

Molds

1. Obtain the plate culture of *Rhizopus*. **Do not remove the lid. Uncovering the organism will spread spores and contaminate the laboratory.**
   a. Examine the colony morphology and sketch a representative colony. Record the **color** on both the front (obverse) and reverse surfaces. Also record the **colony texture** as glabrous (leathery), velvety, yeast like, cottony, or granular (powdery), and the **colony topography** as flat, rugose (with radial grooves), folded, crateriform, verrucose (warty, rough), or cerebriform (brainlike).
   b. Examine the colony under the dissecting microscope and identify hyphae, rhizoids, and sporangia. Sketch and label representative structures.
2. Examine prepared slides of *Rhizopus* sporangia using medium and high dry powers. Identify the following: sporangiophores, sporangia, and spores. Sketch and label representative structures.
3. Examine prepared slides of *Rhizopus* gametangia using medium and high dry power. Identify the following: progametangia, gametangia, young zygosporangia, mature zygosporangia. Sketch and label representative structures.
4. Obtain the plate culture of *Penicillium*. **Do not remove the plate's lid or you will spread spores and contaminate the laboratory.**
   a. Examine the colony morphology and sketch a representative colony. Record the **color** on both the front (obverse) and reverse surfaces. Also record the **colony texture** as glabrous (leathery), velvety, yeast like, cottony, or granular (powdery), and the **colony topography** as flat, rugose (with radial grooves), folded, crateriform, verrucose (warty, rough), or cerebriform (brainlike).
   b. Examine the colony under the dissecting microscope. Identify hyphae and conidia. Sketch and label representative structures.
5. Observe prepared slides of *Penicillium* conidiophores. Identify the following: hyphae, conidiophores, and chains of conidia. Sketch and label representative structures.
6. Obtain the plate culture of *Aspergillus*. **Do not remove the plate's lid or you will spread spores and contaminate the laboratory.**
   a. Examine the colony morphology and sketch a representative colony. Record the **color** on both the front (obverse) and reverse surfaces. Also record the **colony texture** as glabrous (leathery), velvety, yeast like, cottony, or granular (powdery), and the **colony topography** as flat, rugose (with radial grooves), folded, crateriform, verrucose (warty, rough), or cerebriform (brainlike).
   b. Examine the colony under the dissecting microscope. Identify hyphae and conidia. Sketch and label representative structures.
7. Observe prepared slides of *Aspergillus* conidiophores. Identify hyphae, conidiophores, and conidia. Sketch and label representative structures.

REFERENCES

Collins, C. H., Patricia M. Lyne, and J. M. Grange. 1995. Chapter 51 in *Collins and Lyne's Microbiological Methods*, 7th Ed. Butterworth-Heineman, Oxford.

Fisher, Fran and Norma B. Cook. 1998. Chapter 2 in *Fundamentals of Diagnostic Mycology*. W. B. Saunders Company, Philadelphia, PA.

Forbes, Betty A., Daniel F. Sahm, and Alice. S. Weissfeld. 2002. Chapter 53 in *Bailey and Scott's Diagnostic Microbiology*, 11th Ed. Mosby-Yearbook, Inc., St. Louis, MO.

Koneman, Elmer W., Stephen D. Allen, William M. Janda, Paul C. Schreckenberger, and Washington C. Winn, Jr. 1997. Chapter 19 in *Color Atlas and Textbook of Diagnostic* Microbiology, 5th Ed. J. B. Lippincott Company, Philadelphia, PA.

10-2 PROTOZOANS

Protozoans are unicellular eukaryotic heterotrophic microorganisms belonging to the Kingdom Protista. Protozoans are further classified as follows: Phylum Sarcomastigophora (including Subphylum Mastigophora [the flagellates] and Subphylum Sarcodina [the amebas]), Phylum Ciliophora (the ciliates), and Phylum Apicomplexa (sporozoans and others). A typical life cycle includes a vegetative **trophozoite** and a resting **cyst** stage. Some have additional stages, making their life cycles more complex.

Photographic Atlas Reference
Protozoan Survey Page 175
Protozoans of Clinical Importance Page 176

MATERIAL NEEDED FOR THIS EXERCISE

Per Student Group

- Fresh culture of *Amoeba spp.*
- Fresh culture of *Paramecium spp.*
- Fresh culture of *Euglena spp.*
- Methyl cellulose
- Clean microscope slides and cover glasses
- Methylene blue stain
- Prepared slides of:
  - *Entamoeba histolytica* trophozoite and cyst
  - *Entamoeba coli* trophozoite and cyst
  - *Balantidium coli* trophozoite and cyst
  - *Giardia lamblia* trophozoite and cyst
  - *Trichomonas vaginalis* trophozoite
  - *Trypanosoma spp.*
  - *Plasmodium spp.*
  - *Toxoplasma gondii* trophozoite

PROCEDURE

1. Make wet mount preparations of the living specimens as illustrated in Figure 3-6. You may wish to add a drop of methylene blue to stain the organisms. (If you don't stain them, reduce the light using the iris diaphragm to improve contrast.) A drop of methyl cellulose may also be useful as this slows down the fast swimmers. Observe the following structures on each organism. Sketch and label these.

Amoeba	*Paramecium*	*Euglena*
nucleus	macronucleus	nucleus
pseudopods	micronucleus	flagellum
vacuoles	cilia	
	oral groove	
	contractile vacuole	

2. Obtain prepared slides of the protozoan pathogens and observe them under appropriate magnification. You should observe the assigned structures on each organism. (Many of these slides are made from patient samples, so there will be a lot of other material on the slide besides the desired organism. You must search carefully and with patience.) Sketch and label these.

Entamoeba histolytica
- Trophozoite
  - pseudopods
  - nucleus with small, central karyosome and beaded nucleus
  - ingested erythrocytes
- Cyst
  - multiple nuclei (up to four) with karyosomes and chromatin as in the trophozoite
  - cytoplasmic chromatoidal bars (maybe)

Entamoeba coli
- Trophozoite
  - same as *E. histolytica* except with eccentric karyosome and unclumped chromatin
- Cyst
  - up to 8 nuclei (more than 4 is enough to distinguish it from *E. histolytica*) that are the same as in the trophozoite
  - cytoplasmic chromatoidal bars (maybe)

Balantidium coli
- Trophozoite
  - elongated shape
  - cilia
  - macronucleus
  - micronucleus (maybe)
- Cyst
  - spherical shape with multiple nuclei

Giardia lamblia
- Trophozoite
  - oval shape
  - flagella (four pairs)
  - nuclei (two)
  - median bodies (two)
- Cyst
  - multiple nuclei (four)
  - median bodies (four)

Trichomonas vaginalis trophozoite
- nucleus
- flagella (four)

Trypanosoma spp.
- nucleus
- flagellum
- undulating membrane

Plasmodium falciparum
- ring stage
- mature trophozoite
- schizont
- male gametocyte
- female gametocyte

Toxoplasma gondii
- trophozoite
- bow-shaped cells
- nucleus

REFERENCES

Ash, Lawrence R. and Thomas C. Orihel. 1991. Chapter 14 in *Parasites: A Guide to Laboratory Procedures and Identification.* American Society for Clinical Pathology (ASCP) Press, Chicago, IL.

Forbes, Betty A., Daniel F. Sahm, and Alice. S. Weissfeld. 2002. Chapter 52 (pages 650–687) in *Bailey and Scott's Diagnostic Microbiology*, 11th Ed. Mosby-Yearbook, Inc., St. Louis, MO.

Garcia, Lynne Shore. 2001. Chapters 2, 3, 5, 7 and 9 in *Diagnostic Medical Parasitology*, 4th Ed. ASM Press, Washington, DC.

Koneman, Elmer W., Stephen D. Allen, William M. Janda, Paul C. Schreckenberger, and Washington C. Winn, Jr. 1997. Chapter 20 in *Color Atlas and Textbook of Diagnostic Microbiology*, 5th Ed. J. B. Lippincott Company, Philadelphia, PA.

Lee, John J., Seymour H. Hutner, and Eugene C. Bovee. 1985. *Illustrated Guide to the Protozoa.* Society of Protozoologists, Lawrence, KS.

Markell, Edward K., Marietta Voge, and David T. John. 1992. *Medical Parasitology*, 7th Ed. W. B. Saunders Company, Philadelphia, PA.

10-3 HELMINTH PARASITES

Photographic Atlas Reference
Helminths Page 187

A study of helminths is appropriate to the microbiology lab because clinical specimens may contain microscopic evidence of helminth infection. The three major groups of parasitic worms encountered in lab situations are the trematodes (flukes), the cestodes (tapeworms), and the nematodes (round worms). Life cycles of the parasitic worms are often complex, sometimes involving several hosts, and are beyond the scope of this book. Emphasis here is on clinically important diagnostic features of each worm.

MATERIALS NEEDED FOR THIS EXERCISE

- Prepared slides of:
  - *Ascaris lumbricoides* eggs in a fecal smear
  - *Clonorchis sinensis* in a fecal smear
  - *Dipylidium caninum* eggs in a fecal smear
  - *Echinococcus granulosus* hydatid cyst in section
  - *Enterobius vermicularis* eggs in a fecal smear
  - Hookworm *Ancylostoma duodenale* or *Necator americanus* eggs in a fecal smear
  - *Hymenolepis nana* eggs in a fecal smear
  - *Paragonimus westermani* eggs in a fecal smear
  - *Schistosoma mansoni* eggs in a fecal smear
  - *Strongyloides stercoralis* rhabditiform larva in a fecal smear
  - *Taenia solium* proglottid (whole mount)
  - *Taenia solium* scolex (whole mount)
  - *Taenia spp.* eggs in a fecal smear
  - *Wuchereria bancrofti* microfilariae in a blood smear

PROCEDURE

1. Observe the prepared slides provided of the helminth specimens in fecal smears and other tissues. Scanning on low power (10X objective) is best for most preparations, then move to high dry or oil immersion to see detail. Most egg specimens are in fecal smears, so there will be a lot of other material on the slide besides the eggs. You must search carefully and with patience. Sketch what you see and measure dimensions. Be able to identify each to species if shown an unlabeled specimen.

REFERENCES

Ash, Lawrence R. and Thomas C. Orihel. 1991. Chapters 15 and 16 in *Parasites: A Guide to Laboratory Procedures and Identification.* American Society for Clinical Pathology (ASCP) Press, Chicago, IL.

Forbes, Betty A., Daniel F. Sahm, and Alice. S. Weissfeld. 2002. Chapter 52 in (Pages 687–698) in *Bailey and Scott's Diagnostic Microbiology*, 11th Ed. Mosby-Yearbook, Inc., St. Louis, MO.

Garcia, Lynne Shore. 2001. Chapters 10, 12, 13, 14, 16, and 17 in *Diagnostic Medical Parasitology*, 4th Ed. ASM Press, Washington, DC.

Koneman, Elmer W., Stephen D. Allen, William M. Janda, Paul C. Schreckenberger, and Washington C. Winn, Jr. 1997. Chapter 20 in *Color Atlas and Textbook of Diagnostic Microbiology*, 5th Ed. J. B. Lippincott Company, Philadelphia, PA.

Orihel, Thomas C., and Lawrence R. Ash. 1999. Chapter 112 in *Manual of Clinical Microbiology*, 7th Ed., edited by Patrick R. Murray, Ellen Jo Baron, Michael A. Pfaller, Fred C. Tenover, and Robert H. Yolken. American Society for Microbiology, Washington, DC.

Roberts, Larry S. and John Janovy, Jr. *Foundations of Parasitology,* 5th Ed. Wm. C. Brown Publishers, Dubuque, IA.

APPENDIX A

Media, Reagent, and Stain Recipes

Media

Bile Esculin Agar

Beef extract	3.0 g
Peptone	5.0 g
Oxgall	40.0 g
Esculin	1.0 g
Ferric citrate	0.5 g
Agar	15.0 g
Distilled or deionized water	1.0 L

pH 6.4 – 6.8 at 25°C

1. Suspend the ingredients in one liter of distilled or deionized water, mix well, and boil to dissolve completely.
2. Dispense 7.0 mL volumes into test tubes and cap loosely.
3. Sterilize in the autoclave at 15 lbs. pressure (121°C) for 15 minutes.
4. Remove from the autoclave, slant, and allow the medium to cool to room temperature.

Blood Agar

Infusion from beef heart (solids)	2.0 g
Pancreatic digest of casein	13.0 g
Sodium chloride	5.0 g
Yeast extract	5.0 g
Agar	15.0 g
Defibrinated sheep blood	50.0 mL
Distilled or deionized water	1.0 L

pH 7.1–7.5 at 25°C

1. Suspend, mix, and boil the dry ingredients in one liter distilled or deionized water. This is blood agar base.
2. Cover loosely and sterilize in the autoclave at 15 lbs. pressure (121°C) for 15 minutes.
3. Remove from the autoclave and cool to 45°C.
4. Aseptically add the sterile, room temperature sheep blood to the blood agar base and mix well.
5. Pour into sterile Petri dishes and allow the medium to cool to room temperature.

Brilliant Green Lactose Bile Broth

Peptone	10.0 g
Lactose	10.0 g
Oxgall	20.0 g
Brilliant green dye	0.0133 g
Distilled or deionized water	1.0 L

pH 7.0–7.4 at 25°C

1. Suspend, mix, and heat the ingredients in one liter of distilled or deionized water until completely dissolved.
2. Dispense 10.0 mL portions into test tubes.
3. Place an inverted Durham tube in each broth and cap loosely.
4. Sterilize in the autoclave at 15 lbs. pressure (121°C) for 15 minutes.
5. Remove the medium from the autoclave and allow it to cool before inoculating.

Citrate Agar (Simmons)

Ammonium dihydrogen phosphate	1.0 g
Dipotassium phosphate	1.0 g
Sodium chloride	5.0 g
Sodium citrate	2.0 g
Magnesium sulfate	0.2 g
Agar	15.0 g
Bromthymol blue	0.08 g
Distilled or deionized water	1.0 L

pH 6.7–7.1 at 25°C

1. Suspend the ingredients in one liter distilled or deionized water, mix well, and boil to dissolve completely.
2. Dispense 7.0 mL portions into test tubes and cap loosely.
3. Sterilize in the autoclave at 15 lbs pressure (121°C) for 15 minutes.
4. Remove from the autoclave, slant, and cool to room temperature.

Complete Medium (Ames Test)

Beef extract	3.0 g
Peptone	5.0 g
Sodium chloride	5.0 g
Agar	20.0 g
Distilled or deionized water	1.0 L

1. Suspend, mix, and boil the ingredients in one liter of distilled or deionized water until completely dissolved.
2. Cover loosely and sterilize in the autoclave at 15 lbs. pressure (121°C) for 15 minutes.
3. Remove from the autoclave and cool slightly.
4. Aseptically pour into sterile petri dishes (20 mL/plate) and cool to room temperature.

Decarboxylase Medium (Møller)

Peptone	5.0 g
Beef extract	5.0 g
Glucose (dextrose)	0.5 g
Bromcresol purple	0.01 g
Cresol red	0.005 g
Pyridoxal	0.005 g
L-Lysine, L-Ornithine, or L-Arginine	10.0 g
Distilled or deionized water	1.0 L

pH 5.8–6.2 at 25°C

1. Suspend and heat the ingredients in one liter of distilled or deionized water until completely dissolved. (Use only one of the listed L-amino acids.)
2. Adjust pH by adding NaOH if necessary.
3. Dispense 7.0 mL volumes into test tubes and cap.
4. Sterilize in the autoclave at 15 lbs. pressure (121°C) for 10 minutes.
5. Remove from the autoclave and cool to room temperature.

Desoxycholate Agar (Modified Leifson)

Peptone	10.0 g
Lactose	10.0 g
Sodium desoxycholate	1.0 g
Sodium chloride	5.0 g
Dipotassium phosphate	2.0 g
Ferric citrate	1.0 g
Sodium citrate	1.0 g
Agar	16.0 g
Neutral red	0.033 g
Distilled or deionized water	1.0 L

pH 7.1–7.5 at 25°C

1. Heat and stir the ingredients in one liter of distilled or deionized water. Boil for 1 minute to make certain they are completely dissolved.
2. When cooled to 50°C, pour into sterile plates.
3. Allow the medium to cool to room temperature.

DNase Test Agar with Methyl Green

Tryptose	20.0 g
Deoxyribonucleic acid	2.0 g
Sodium chloride	5.0 g
Agar	15.0 g
Methyl green	0.05 g
Distilled or deionized water	1.0 L

pH 7.1–7.5 at 25°C

1. Suspend, mix, and boil the ingredients in one liter of distilled or deionized water until completely dissolved.
2. Cover loosely and sterilize in the autoclave at 15 lbs. pressure (121°C) for 15 minutes.
3. Aseptically pour into sterile Petri dishes (20 mL/plate) and cool to room temperature.

EC Broth

Tryptose	20.0 g
Lactose	5.0 g
Dipotassium phosphate	4.0 g
Monopotassium phosphate	1.5 g
Sodium chloride	5.0 g
Distilled or deionized water	1.0 L

pH 6.7–7.1 at 25°C

1. Suspend, mix, and heat the ingredients in one liter of distilled or deionized water until completely dissolved.
2. Dispense 10.0 mL portions into test tubes.
3. Place an inverted Durham tube in each broth and cap loosely.
4. Sterilize in the autoclave at 15 lbs. pressure (121°C) for 15 minutes.
5. Remove the media from the autoclave and allow it to cool before inoculating.

Endo Agar

Peptone	10.0 g
Lactose	10.0 g
Dipotassium phosphate	3.5 g
Agar	15.0 g
Basic Fuchsin	0.5 g
Sodium Sulfite	2.5 g
Distilled or deionized water	1.0 L

pH 7.3–7.7 at 25°C

1. Mix and heat the ingredients in one liter of distilled or deionized water until they are completely dissolved.
2. Autoclave for 15 minutes at 15 lbs. pressure (121°C).
3. When cooled to 50°C, pour into sterile plates.
4. Allow the medium to cool to room temperature.

Enriched TSA (See Tryptic Soy Agar)

Eosin Methylene Blue Agar (Levine)

Peptone	10.0 g
Lactose	10.0 g*
Dipotassium phosphate	2.0 g
Agar	15.0 g
Eosin Y	0.4 g
Methylene blue	0.065 g
Distilled or deionized water	1.0 L

pH 6.9–7.3 at 25°C

1. Mix and heat the ingredients in one liter of distilled or deionized water until they are completely dissolved.
2. Autoclave for 15 minutes at 15 lbs. pressure (121°C).
3. When cooled to 50°C, pour into sterile plates.
4. Allow the medium to cool to room temperature.

Glucose Broth

Peptone	10.0 g
Glucose	5.0 g
NaCl	5.0 g
Distilled or deionized water	1.0 L

1. Suspend the ingredients in one liter of distilled or deionized water. Agitate and heat slightly (if necessary) to dissolve completely.
2. Dispense 7.0 mL portions into test tubes and cap loosely.
3. Autoclave for 15 minutes at 121°C to sterilize the medium.

Glucose Salts Medium

Glucose	5.0 g
NaCl	5.0 g
$MgSO_4$	0.2 g
$(NH_4)H_2PO_4$	1.0 g
K_2HPO_4	1.0 g
Distilled or deionized water	1.0 L

1. Suspend the ingredients in one liter of distilled or deionized water. Agitate and heat slightly (if necessary) to dissolve completely.
2. Dispense 7.0 mL portions into test tubes and cap loosely.
3. Autoclave for 15 minutes at 121°C to sterilize the medium.

Glycerol Yeast Extract Agar

Glycerol	5.0 mL
Yeast extract	2.0 g
Dipotassium phosphate	1.0 g
Agar	15.0 g
Distilled or deionized water	1.0 L

1. Mix and heat the ingredients in one liter of distilled or deionized water until they are completely dissolved.
2. Autoclave for 15 minutes at 15 lbs. pressure (121°C).
3. When cooled to 50°C, pour into sterile plates.
4. Allow the medium to cool to room temperature.

* An alternate recipe replaces the 10.0 g of lactose with 5.0 g of lactose and 5.0 g of sucrose.

Hektoen Enteric Agar

Yeast extract	3.0 g
Peptic digest of animal tissue	12.0 g
Lactose	12.0 g
Sucrose	12.0 g
Salicin	2.0 g
Bile salts	9.0 g
Sodium chloride	5.0 g
Sodium thiosulfate	5.0 g
Ferric ammonium citrate	1.5 g
Bromthymol blue	0.064 g
Acid fuchsin	0.1 g
Agar	13.5 g
Distilled or deionized water	1.0 L

pH 7.4–7.8 at 25°C

1. Mix and heat the ingredients in one liter of distilled or deionized water until they are dissolved. Boil for 1 minute to make certain they are completely dissolved.
2. Do not autoclave.
3. When cooled to 50°C, pour into sterile plates.
4. Cool to room temperature with lids slightly open.

Kligler's Iron Agar

Beef extract	3.0 g
Yeast extract	3.0 g
Peptone	15.0 g
Proteose peptone	5.0 g
Lactose	10.0 g
Dextrose (glucose)	1.0 g
Ferrous sulfate	0.2 g
Sodium chloride	5.0 g
Sodium thiosulfate	0.3 g
Agar	12.0 g
Phenol red	0.024 g
Distilled or deionized water	1.0 L

pH 7.2–7.6 at 25°C

1. Suspend, mix, and boil the ingredients in one liter of distilled or deionized water until completely dissolved.
2. Transfer 7.0 mL portions to test tubes and cap loosely.
3. Sterilize in the autoclave at 15 lbs. pressure (121°C) for 15 minutes.
4. Remove from the autoclave and slant in such a way as to form a deep butt.
5. Allow the medium to cool to room temperature.

Lauryl Tryptose Broth

Tryptose	20.0 g
Lactose	5.0 g
Dipotassium phosphate	2.75 g
Monopotassium phosphate	2.75 g
Sodium chloride	5.0 g
Sodium lauryl sulfate	0.1 g
Distilled or deionized water	1.0 L

pH 6.6–7.0 at 25°C

1. Suspend, mix, and heat the ingredients in one liter of distilled or deionized water until completely dissolved.
2. Dispense 10.0 mL portions into test tubes.
3. Place an inverted Durham tube in each broth and cap loosely.
4. Sterilize in the autoclave at 15 lbs. pressure (121°C) for 15 minutes.
5. Remove the medium from the autoclave and allow it to cool before inoculating.

Litmus Milk Medium

Skim milk	100.0 g
Azolitmin	0.5 g
Sodium sulfite	0.5 g
Distilled or deionized water	1.0 L

pH 6.3–6.7 at 25°C

1. Suspend and mix the ingredients in one liter of deionized or distilled water and heat to approximately 50°C to dissolve completely.
2. Transfer 7.0 mL portions to test tubes and cap loosely.
3. Sterilize in the autoclave at 113–115°C for 20 minutes.
4. Remove from the autoclave and allow the medium to cool to room temperature.

Luria-Bertani Broth

Bacto-tryptone	10.0 g
Bacto-yeast Extract	5.0 g
NaCl	10.0 g
Distilled or deionized water	1.0 L

pH 7.4 at 25°C

1. Suspend the ingredients in one liter of distilled or deionized water. Agitate and heat slightly (if necessary) to dissolve completely.
2. Dispense 7.0 mL portions into test tubes and cap loosely.
3. Autoclave for 15 minutes at 121°C to sterilize the medium.

Lysine Iron Agar

Peptone	5.0 g
Yeast Extract	3.0 g
Dextrose	1.0 g
L-Lysine hydrochloride	10.0 g
Ferric ammonium citrate	0.5 g
Sodium thiosulfate	0.04 g
Bromcresol purple	0.02 g
Agar	15.0 g

pH 6.5–6.9 at 25°C

1. Suspend and mix the ingredients in one liter of deionized or distilled water and boil to completely dissolve.
2. Transfer 8.0 mL portions to test tubes and cap loosely.
3. Sterilize in the autoclave at 121°C for 15 minutes.
4. Remove from the autoclave and slant in such a way as to produce a deep butt. Allow the medium to cool to room temperature.

MacConkey Agar

Pancreatic digest of gelatin	17.0 g
Pancreatic digest of casein	1.5 g
Peptic digest of animal tissue	1.5 g
Lactose	10.0 g
Bile salts	1.5 g
Sodium chloride	5.0 g
Neutral red	0.03 g
Crystal violet	0.001 g
Agar	13.5 g
Distilled or deionized water	1.0 L

pH 6.9–7.3 at 25°C

1. Mix and heat the ingredients in one liter of distilled or deionized water until they are dissolved. Boil for 1 minute to make certain they are completely dissolved.
2. Autoclave for 15 minutes at 15 lbs. pressure (121°C).
3. When cooled to 50°C, pour into sterile plates.
4. Allow the medium to cool to room temperature.

Malonate Broth

Yeast extract	1.0 g
Ammonium sulfate	2.0 g
Dipotassium phosphate	0.6 g
Monopotassium phosphate	0.4 g
Sodium chloride	2.0 g
Sodium malonate	3.0 g
Dextrose	0.25 g
Bromthymol blue	0.025 g
Distilled or deionized water	1.0 L

pH 6.5–6.9 at 25°C

1. Suspend the ingredients in one liter of distilled or deionized water. Agitate and heat slightly (if necessary) to dissolve completely.
2. Dispense 7.0 mL portions into test tubes and cap loosely.
3. Autoclave for 15 minutes at 121°C to sterilize the medium.

Mannitol Salt Agar

Beef extract	1.0 g
Peptone	10.0 g
Sodium chloride	75.0 g
D-Mannitol	10.0 g
Phenol red	0.025 g
Agar	15.0 g
Distilled or deionized water	1.0 L

pH 7.2–7.6 at 25°C

1. Suspend the ingredients in one liter of distilled or deionized water and mix. Boil one minute to completely dissolve ingredients.
2. Autoclave for 15 minutes at 15 lbs. pressure (121°C).
3. When cooled to 50°C, pour into sterile plates.
4. Allow the medium to cool to room temperature.

Milk Agar

Beef extract	3.0 g
Peptone	5.0 g
Agar	15.0 g
Powdered nonfat milk	100.0 g
Distilled or deionized water	1.0 L

pH 7.0–7.4 at 25°C

1. Suspend the powdered milk in 500.0 mL of distilled or deionized water in a one-liter flask, mix well, and cover loosely.
2. Suspend the remainder of the ingredients in 500.0 mL of deionized water in a one liter flask, mix well, boil to dissolve completely, and cover loosely.
3. Sterilize in the autoclave at 113–115°C for 20 minutes.
4. Remove from the autoclave, allow to cool slightly, then aseptically pour the milk solution into the agar solution and mix *gently* (to prevent foaming).
5. Aseptically pour into sterile Petri dishes (15 mL/plate).
6. Allow the medium to cool to room temperature.

Minimal Medium (Ames Test)

Dextrose (glucose)	20.0 g
50x Vogel-Bonner salts	20.0 mL
Histidine	0.00016 g
Biotin	0.00025 g
Agar	20.0 g
Distilled or deionized water	1.0 L

1. Add 1.6 mg histidine to 10.0 mL distilled or deionized water and filter sterilize.
2. Add 2.5 mg biotin to 10.0 mL distilled or deionized water and filter sterilize.
3. Prepare the 50x Vogel-Bonner salts solution by adding the ingredients to *just enough* water to dissolve them while heating and stirring. After the ingredients are dissolved add enough water to bring the total volume up to exactly one liter.
4. Suspend, mix, and boil the agar in 500.0 mL of distilled or deionized water until completely dissolved.
5. Suspend and mix the dextrose in 500.0 mL of distilled or deionized water until completely dissolved.
6. Cover the agar and dextrose containers loosely and sterilize in the autoclave at 121°C for 15 minutes.
7. Remove from the autoclave and allow to cool to 80°C.
8. Aseptically add 1.0 mL histidine solution, 1.0 mL biotin solution, and 20 mL 50x Vogel-Bonner salts to the glucose solution and mix well.
9. Add the glucose solution to the agar solution, mix well and aseptically pour into sterile Petri dishes (20 mL/plate). Allow the medium to cool to room temperature.

Motility Test Medium

Beef extract	3.0 g
Pancreatic digest of gelatin	10.0 g
Sodium chloride	5.0 g
Agar	4.0 g
Triphenyltetrazolium chloride (TTC)	0.05 g
Distilled or deionized water	1.0 L

pH 7.1–7.5 at 25°C

1. Suspend the ingredients in one liter of distilled or deionized water, mix well, and boil to dissolve completely.
2. Dispense 7.0 mL portions into test tubes and cap loosely.
3. Sterilize in the autoclave at 15 lbs. pressure (121°C) for 15 minutes.
4. Remove from the autoclave and allow it to cool in the upright position.

MRVP Broth

Buffered peptone	7.0 g
Dipotassium phosphate	5.0 g
Dextrose (glucose)	5.0 g
Distilled or deionized water	1.0 L

pH 6.7–7.1 at 25°C

1. Suspend the ingredients in one liter of deionized or distilled water, mix well, and warm until completely dissolved.
2. Transfer 7.0 mL portions to test tubes and cap loosely.
3. Sterilize in the autoclave at 15 lbs. pressure (121°C) for 15 minutes.
4. Remove from the autoclave and allow the medium to cool to room temperature.

Mueller-Hinton II Agar

Beef extract	2.0 g
Acid hydrolysate of casein	17.5 g
Starch	1.5 g
Agar	17.0 g
Distilled or deionized water	1.0 L

pH 7.1–7.5 at 25°C

1. Suspend the ingredients in one liter of distilled or deionized water, mix well and boil to dissolve completely.
2. Cover loosely and sterilize in the autoclave at 121°C (15 lbs.) for 15 minutes.
3. Remove from the autoclave, allow to cool slightly.
4. Aseptically pour into sterile Petri dishes to a depth of 4 mm.
5. Allow the medium to cool to room temperature.

Nutrient Agar

Beef extract	3.0 g
Peptone	5.0 g
Agar	15.0 g
Distilled or deionized water	1.0 L

pH 6.6–7.0 at 25°C

Plates

1. Suspend the ingredients in one liter of distilled or deionized water, mix well, and boil to dissolve completely.
2. Cover loosely and sterilize in the autoclave at 15 lbs. pressure (121°C) for 15 minutes.
3. Remove from the autoclave, allow to cool slightly and aseptically pour into sterile Petri dishes (20 mL/plate).
4. Allow the medium to cool to room temperature.

Tubes

1. Suspend the ingredients in one liter of distilled or deionized water, mix well, and boil until fully dissolved.
2. Dispense 10 mL portions into test tubes and cap loosely.
3. Autoclave for 15 minutes at 121°C to sterilize the medium.
4. Cool to room temperature with the tubes in an upright position for agar deep tubes. Cool with the tubes on an angle for agar slants.

Nutrient Broth

Beef extract	3.0 g
Peptone	5.0 g
Distilled or deionized water	1.0 L

pH 6.6–7.0 at 25°C

1. Suspend the ingredients in one liter of distilled or deionized water. Agitate and heat slightly (if necessary) to dissolve completely.
2. Dispense 7.0 mL portions into test tubes and cap loosely.
3. Autoclave for 15 minutes at 121°C to sterilize the medium.

Nutrient Gelatin

Beef Extract	3.0 g
Peptone	5.0 g
Gelatin	120.0 g
Distilled or deionized water	1.0 L

pH 6.6–7.0 at 25°C

1. Slowly add the ingredients to one liter of distilled or deionized water while stirring.
2. Warm to >50°C and maintain temperature until completely dissolved.
3. Dispense 7.0 mL volumes into test tubes and cap loosely.
4. Sterilize in the autoclave at 15 lbs. pressure (121°C) for 15 minutes.
5. Remove from the autoclave immediately and allow the medium to cool to room temperature in the upright position.

OF Basal Medium (see OF Carbohydrate Solution)

Pancreatic digest of casein	2.0 g
Sodium chloride	5.0 g
Dipotassium phosphate	0.3 g
Agar	2.5 g
Bromthymol blue	0.03 g
Distilled or deionized water	1.0 L

pH 6.6–7.0 at 25°C

OF Carbohydrate Solution (see OF Basal Medium)

Carbohydrate (glucose, lactose, sucrose)	1.0 g
Distilled or deionized water to total	10.0 mL

1. Suspend the ingredients, *without the carbohydrate*, in one liter of distilled or deionized water, mix well, and boil to dissolve completely. This is basal medium.
2. Divide the medium into ten aliquots of 100.0 mL each.
3. Cover loosely and sterilize in the autoclave at 121°C for 15 minutes.
4. Prepare carbohydrate solution, cover loosely and autoclave at 118°C for 10 minutes.
5. Allow both solutions to cool to 50°C.
6. Aseptically add 10.0 mL sterile carbohydrate solution to a basal medium aliquot and mix well.
7. Aseptically transfer 7.0 mL volumes to sterile test tubes and allow to cool.

Phenol Red (Carbohydrate) Broth

Pancreatic digest of casein	10.0 g
Sodium chloride	5.0 g
Carbohydrate (glucose, lactose, sucrose)	5.0 g
Phenol red	0.018 g
Distilled or deionized water	1.0 L

pH 7.1 ± 7.5 at 25°C

1. Suspend the ingredients in one liter of distilled or deionized water, mix well, and warm slightly to dissolve completely.
2. Dispense 7.0 mL volumes into test tubes.
3. Insert inverted Durham tubes into the test tubes and cap loosely.
4. Sterilize in the autoclave at 116–118°C for 15 minutes.
5. Remove from the autoclave and allow the medium to cool to room temperature.

Phenylalanine Deaminase Agar

DL-Phenylalanine	2.0 g
Yeast extract	3.0 g
Sodium chloride	5.0 g
Sodium phosphate	1.0 g
Agar	12.0 g
Distilled or deionized water	1.0 L

pH 7.1–7.5 at 25°C

1. Suspend the ingredients in one liter of distilled or deionized water, mix well, and boil to dissolve completely.
2. Dispense 7.0 mL volumes into test tubes and cap loosely.
3. Sterilize in the autoclave at 15 lbs pressure (121°C) for 10 minutes.
4. Remove from the autoclave, slant, and allow the medium to cool to room temperature.

Phenylethyl Alcohol Agar

Tryptose	10.0 g
Beef extract	3.0 g
Sodium chloride	5.0 g
Phenylethyl alcohol	2.5 g
Agar	15.0 g
Distilled or deionized water	1.0 L

pH 7.1–7.5 at 25°C

1. Suspend the ingredients in one liter of distilled or deionized water. Boil one minute to completely dissolve ingredients.
2. Autoclave for 15 minutes at 15 lbs. pressure (121°C).
3. When cooled to 50°C, pour into sterile plates.
4. Allow the medium to cool to room temperature.

Photobacterium Broth

Tryptone	5.0 g
Yeast extract	2.5 g
Ammonium chloride	0.3 g
Magnesium sulfate	0.3 g
Ferric chloride	0.01 g
Calcium carbonate	1.0 g
Monopotassium phosphate	3.0 g
Sodium glycerol phosphate	23.5 g
Sodium chloride	30.0 g
Distilled or deionized water	1.0 L

pH 6.8–7.2 at 25°C

1. Suspend the ingredients in one liter of distilled or deionized water. Boil one minute to completely dissolve ingredients.
2. Dispense into flasks to form a shallow layer.
3. Autoclave for 15 minutes at 15 lbs. pressure (121°C).
4. Allow the medium to cool to room temperature.

Potato Dextrose Agar

Potato flakes	20.0 g
Dextrose	10.0 g
Agar	15.0 g
Distilled or deionized water	1.0 L

1. Mix and boil the ingredients in one liter of distilled or deionized water until they are completely dissolved.
2. Autoclave for 15 minutes at 15 lbs. pressure (121°C).
3. Remove the agar mixture from the autoclave and cool to 50°C.
4. Mix and pour into sterile Petri dishes. Note: Swirl the flask frequently to keep the flakes uniformly distributed.

Purple Broth

Peptone	10.0 g
Beef extract	1.0 g
Sodium chloride	5.0 g
Bromcresol purple	0.02 g
Carbohydrate (glucose, lactose, or sucrose)	10.0 g
Distilled or deionized water	1.0 L

pH 6.6–7.0 at 25°C

1. Suspend the ingredients in one liter of distilled or deionized water, mix well, and warm slightly to dissolve completely.
2. Dispense 9.0 mL volumes into test tubes.
3. Insert inverted Durham tubes into the test tubes and cap loosely.
4. Sterilize in the autoclave at 118°C for 15 minutes.
5. Remove from the autoclave and allow the medium to cool to room temperature.

Sabouraud Dextrose Agar (with antibiotics added to inhibit bacterial growth)

Peptone	10.0 g
Dextrose	40.0 g
Agar	15.0 g
Penicillin*	20000.0 units
Streptomycin*	0.00004 g
Distilled or deionized water	1.0 L

pH 5.2–5.6 at 25°C

1. Mix and heat the ingredients in one liter of distilled or deionized water until they are completely dissolved.

---

* To obtain the desired proportions of antibiotics in the medium, prepare as follows:
1. Dissolve 100,000 units penicillin in 10 mL sterile distilled or deionized water. Add 2 mL to 1.0 liter of agar medium.
2. Dissolve 1.0 g streptomycin in 10 mL sterile distilled or deionized water. Add 1.0 mL of this mixture to 9.0 mL sterile distilled or deionized water. Add 4 mL of this diluted mixture to 1.0 liter of agar medium.

2. Autoclave for 15 minutes at 15 lbs. pressure (121°C).
3. Remove the agar mixture from the autoclave and cool to 50°C.
4. Aseptically add antibiotics. Mix and pour into sterile Petri dishes.
5. Allow the medium to cool to room temperature.

Saline Agar—Double Gel Immunodiffusion

Sodium chloride	10.0 g
Agar	20.0 g
Distilled or deionized water	1.0 L

1. Suspend the ingredients in one liter distilled or deionized water, mix well, and boil to dissolve completely.
2. Pour into Petri dishes to a depth of 3 mm. Do not replace the lids until the agar has solidified and cooled to room temperature.

SIM (Sulfur-Indole-Motility) Medium

Pancreatic digest of casein	20.0 g
Peptic digest of animal tissue	6.1 g
Ferrous ammonium sulfate	0.2 g
Sodium thiosulfate	0.2 g
Agar	3.5 g
Distilled or deionized water	1.0 L

pH 7.1–7.5 at 25°C

1. Suspend the ingredients in one liter of distilled or deionized water, mix well, and boil to dissolve completely.
2. Dispense 7.0 mL volumes into test tubes and cap loosely.
3. Sterilize in the autoclave at 15 lbs pressure (121°C) for 15 minutes.
4. Remove from the autoclave and allow the medium to cool to room temperature.

Snyder Test Medium

Pancreatic digest of casein	13.5 g
Yeast extract	6.5 g
Dextrose	20.0 g
Sodium chloride	5.0 g
Agar	16.0 g
Bromcresol green	0.02 g
Distilled or deionized water	1.0 L

pH 4.6–5.0 at 25°C

1. Suspend the ingredients in one liter distilled or deionized water, mix well, and boil to dissolve completely.
2. Transfer 7.0 mL portions to test tubes and cap loosely.
3. Sterilize in the autoclave at 118–121°C for 15 minutes.
4. Remove from the autoclave and place in a hot water bath set at 45–50°C. Allow at least 30 minutes for the agar temperature to equilibrate before beginning the exercise.

Soft Agar

Beef extract	3.0 g
Peptone	5.0 g
Sodium chloride	5.0 g
Tryptone	2.5 g
Yeast extract	2.5 g
Agar	7.0 g
Distilled or deionized water	1.0 L

1. Suspend, mix and boil the ingredients in one liter of distilled or deionized water until completely dissolved.
2. Transfer 2.5 mL portions to test tubes and cap loosely.
3. Sterilize in the autoclave at 15 lbs. pressure (121°C) for 15 minutes.
4. Remove from the autoclave and place in a hot water bath set at 45°C. Allow 30 minutes for the agar temperature to equilibrate.

Starch Agar

Beef extract	3.0 g
Soluble starch	10.0 g
Agar	12.0 g
Distilled or deionized water	1.0 L

pH 7.3–7.7 at 25°C

1. Suspend the ingredients in one liter of distilled or deionized water, mix well, and boil to dissolve completely.
2. Sterilize in the autoclave at 15 lbs pressure (121°C) for 15 minutes.
3. Remove from the autoclave and allow to cool slightly.
4. Aseptically pour into sterile Petri dishes (20 mL per plate). Allow the medium to cool to room temperature.

Thioglycollate Medium (Fluid)

Yeast Extract	5.0 g
Casitone	15.0 g
Dextrose (glucose)	5.5 g
Sodium chloride	2.5 g
Sodium thioglycolate	0.5 g
L-Cystine	0.5 g
Agar	0.75 g
Resazurin	0.001 g
Distilled or deionized water	1.0 L

pH 7.1–7.5 at 25°C

1. Suspend the ingredients in one liter of distilled or deionized water. Boil to completely dissolve them.
2. Dispense 10.0 mL into sterile test tubes.
3. Autoclave for 15 minutes at 15 lbs. pressure (121°C) to sterilize. Allow the medium to cool to room temperature before inoculating.

Tributyrin Agar

Beef extract	1.5 g
Peptone	2.5 g
Agar	7.5 g
Tributyrin oil	5.0 mL
Distilled or deionized water	500.0 mL

pH 5.8–6.2 at 25°C

1. Suspend the dry ingredients in 500.0 mL of deionized water, mix well, and boil to dissolve completely.
2. Cover loosely and sterilize together with the tube of tributyrin oil in the autoclave at 15 lbs pressure (121°C) for 15 minutes.
3. Remove from the autoclave and aseptically pour agar mixture into a sterile glass blender.
4. Aseptically add the tributyrin oil to the agar mixture and blend on "High" for 1 minute.
5. Aseptically pour into sterile Petri dishes (20 mL/plate). Allow the medium to cool to room temperature.

Triple Sugar Iron Agar

Beef extract	3.0 g
Yeast extract	3.0 g
Peptone	15.0 g
Proteose peptone	5.0 g
Dextrose (glucose)	1.0 g
Lactose	10.0 g
Sucrose	10.0 g
Ferrous sulfate	0.2 g
Sodium chloride	5.0 g
Sodium thiosulfate	0.3 g
Agar	12.0 g
Phenol red	0.024 g
Distilled or deionized water	1.0 L

pH 7.2–7.6 at 25°C

1. Suspend the ingredients in one liter of distilled or deionized water, mix well, and boil to dissolve completely.
2. Transfer 7.0 mL portions to test tubes and cap loosely.
3. Sterilize in the autoclave at 121°C for 15 minutes.
4. Slant in such a way as to form a deep butt.
5. Allow the medium to cool to room temperature.

Tryptic Nitrate Medium

Tryptose	20.0 g
Dextrose	1.0 g
Disodium phosphate	2.0 g
Potassium nitrate	1.0 g
Agar	1.0 g
Distilled or deionized water	1.0 L

pH 7.0–7.4 at 25°C

1. Suspend the ingredients in one liter of deionized or distilled water, mix well and boil until completely dissolved.
2. Transfer 10.0 mL portions to test tubes and cap loosely.
3. Sterilize in the autoclave at 15 lbs. pressure (121°C) for 15 minutes.
4. Remove from the autoclave and allow the medium to cool to room temperature.

Tryptic Soy Agar

Tryptone	15.0 g
Soytone	5.0 g
Sodium Chloride	5.0 g
Agar	15.0 g
Distilled or deionized water	1.0 L

pH 7.1–7.5 at 25°C

Tryptic Soy Agar (Enriched with Yeast Extract)

Tryptone	15.0 g
Soytone	5.0 g
Sodium Chloride	5.0 g
Yeast Extract	5.0 g
Agar	15.0 g
Distilled or deionized water	1.0 L

pH 7.1–7.5 at 25°C

Tryptic Soy Broth

Tryptone	17.0 g
Soytone	3.0 g
Sodium Chloride	5.0 g
Dipotassium phosphate	2.5 g
Distilled or deionized water	1.0 L

pH 7.1–7.5 at 25°C

Urease Agar

Peptone	1.0 g
Dextrose (glucose)	1.0 g
Sodium chloride	5.0 g
Potassium phosphate, monobasic	2.0 g
Agar	15.0 g
Phenol red	0.012 g
Distilled or deionized water	1.0 L

pH 6.6–7.0 at 25°C

1. Suspend the agar in 900 mL distilled or deionized water, mix well, and boil to dissolve completely.
2. Cover loosely and sterilize by autoclaving at 15 lbs. pressure (121°C) for 15 minutes.
3. Remove from the autoclave and allow to cool to 55°C.
4. Suspend the remaining ingredients in 100 mL distilled or deionized water, mix well, and filter sterilize. *Do not autoclave*. This is urease agar base.
5. Aseptically add the urease agar base to the agar solution and mix well.
6. Aseptically transfer 7.0 mL portions to sterile test tubes and cap loosely.
7. Slant in such a way that the agar butt is approximately twice as long as the slant.
8. Allow to cool to room temperature.

Urease Broth

Yeast extract	0.1 g
Potassium phosphate, monobasic	9.1 g
Potassium phosphate, dibasic	9.5 g
Urea	20.0 g
Phenol red	0.01 g
Distilled or deionized water	1.0 L

pH 6.6–7.0 at 25°C

1. Suspend the ingredients in one liter distilled or deionized water and mix well.
2. Filter sterilize the solution. *Do not autoclave.*
3. Aseptically transfer 1.0 mL volumes to small sterile test tubes and cap loosely.

Xylose Lysine Desoxycholate Agar

Xylose	3.5 g
L-Lysine	5.0 g
Lactose	7.5 g
Sucrose	7.5 g
Sodium chloride	5.0 g
Yeast extract	3.0 g
Phenol red	0.08 g
Sodium desoxycholate	2.5 g
Sodium thiosulfate	6.8 g
Ferric ammonium citrate	0.8 g
Agar	13.5 g
Distilled or deionized water	1.0 L

pH 7.3–7.7 at 25°C

1. Suspend the ingredients in one liter of distilled or deionized water and mix. Heat only until the medium boils.
2. Cool in a water bath at 50°C.
3. When cooled, pour into sterile plates.
4. Allow the medium to cool to room temperature.

Yeast Extract Broth

NaCl	5.0 g
Yeast Extract	5.0 g
Nutrient Broth	1.0 L

Reagents

Direct Count—Staining/Diluting agents

See Stains

Lysozyme Buffer

NaCl	22.67 g
Na_2HPO4	3.56 g
KH_2PO4	6.65 g
Distilled or deionized water	1.0 L

1. Dissolve all ingredients in approximately 900 mL water.
2. Add water to bring the total volume up to 1000 mL.

Lysozyme Substrate

Lysozyme Buffer	200.0 mL
Micrococcus lysodeikticus (freeze-dried)	0.1 g

1. Add the freeze-dried *Micrococcus lysodeikticus* to the lysozyme buffer and mix well.
2. Using a spectrophotometer with the wavelength set at 540 μm, adjust the solution's light transmittance to 10% by adding water or *M. lysodeikticus* as needed.

MR/VP (Methyl Red–Voges-Proskauer Test Reagents

Methyl Red

Methyl red dye	0.1 g
Ethanol	300.0 mL
Distilled water to bring volume to	500.0 mL

1. Dissolve the dye in the ethanol.
2. Add water to bring the total volume up to 500 mL.

VP Reagent A

α-naphthol	5.0 g
Absolute Ethanol to bring volume to	100.0 mL

1. Dissolve the α-naphthol in approximately 95 mL of water.
2. Add water to bring the total volume up to 100 mL.

VP Reagent B

Potassium hydroxide	40.0 g
Creatine	0.3 g
Distilled water to bring volume to	100.0 mL

1. Dissolve the potassium hydroxide in approximately 60 mL of water. (**Caution:** This solution is highly concentrated and will become hot as the KOH dissolves. It should be prepared in appropriate glassware on a stirring hot plate.) Allow it to cool to room temperature.
2. Add the creatine. (Creatine enhances the color reaction. As an alternative you can exclude it in the reagent, but add a few crystals to the broth during the test.)
3. Add water to bring the volume up to 100 mL.

McFarland Turbidity Standard (0.5)

Barium chloride ($BaCl_2 \times 2\ H_2O$)	1.175 g
Sulfuric acid, concentrated (H_2SO_4)	1.0 mL
Distilled or deionized water	≅200.0 mL

1. Pour approximately 90 mL of water into a small Erlenmeyer flask.
2. Add the $BaCl_2$ and mix well.
3. Remeasure and add water to bring the total volume up to 100 mL.
4. Add the H_2SO_4 to approximately 90 mL of water.
5. Remeasure and add water to bring the total volume up to 100 mL.
6. Add 0.5 mL of the $BaCl_2$ solution to 99.5 mL of H_2SO_4 and mix well.
7. While keeping the solution well mixed (the barium sulfate will precipitate and settle out) distribute 7 to 10 mL volumes into very clean screw cap test tubes.

Methylene Blue Reductase Reagent

Methylene blue dye	8.8 mg
Distilled or deionized water	200.0 mL

Nitrate Test Reagents

Reagent A

Sulfanilic acid	1.0 g
5N Acetic acid	125.0 mL

Reagent B

Dimethyl-α-naphthylamine	1.0 g
5N Acetic acid	200.0

Oxidase Test Reagent

Tetramethyl-*p*-phenylenediamine dihydrochloride	1.0 g
Deionized water	100.0 mL

Phenylalanine Deaminase Test Reagent

Ferric chloride	10.0 g
Deionized water	≅90.0 mL

1. Dissolve the ferric chloride in approximately 90 mL of distilled or deionized water.
2. Add water to bring the total volume up to 100 mL.

Stains

Acid Fast, Cold Stain Reagents (Modified Kinyoun)

Carbolfuchsin

Basic fuchsin	1.5g
Phenol	4.5 g
Ethanol (95%)	5.0 mL
Isopropanol	20.0 mL
Distilled or deionized water	75.0 mL

1. Dissolve the basic fuchsin in the ethanol and add the isopropanol.
2. Mix/the phenol in the water.
3. Mix the solutions together and let stand for several days.
4. Filter before use.

Decolorizer

H_2SO_4	1.0 mL
Ethanol (95%)	70.0 mL
Distilled or deionized water	29.0 mL

Brilliant Green

Brilliant green dye	1.0 g
Sodium azide	0.01g
Distilled or deionized water	100.0 mL

Acid Fast, Hot Stain Reagents (Ziehl-Neelson)

Carbolfuchsin

Basic fuchsin	0.3 g
Ethanol	10.0 mL
Distilled or deionized water	95.0 mL
Phenol	5.0 mL

1. Dissolve the basic fuchsin in the ethanol.
2. Dissolve the phenol in the water.
3. Combine the solutions and let stand for a few days.
4. Filter before use.

Decolorizer

Ethanol	97.0 mL
HCl (concentrated)	3.0 mL

Methylene Blue Counterstain

Methylene blue chloride	0.3 g
Distilled or deionized water	100.0 mL

CAPSULE STAIN

Congo Red

Congo red dye	5.0 g
Distilled or deionized water	100.0 mL

Maneval's Stain

Phenol (5% aqueous solution)	30.0 mL
Acetic acid, glacial (20% aqueous solution)	10.0 mL
Ferric chloride (30% aqueous solution)	4.0 mL
Acid fuchsin (1% aqueous solution)	2.0 mL

DIRECT COUNT—STAINING/DILUTING AGENTS

Agent A

100% saturated crystal violet-ethanol solution	1.0 mL
NaCl	0.9 g
Distilled or deionized water	≅99.0 mL

1. Dissolve the crystal violet in ethanol and filter.
2. Mix 0.9 g NaCl in approximately 55 mL of distilled or deionized water.
3. Add the NaCl solution to 40 mL of the crystal violet-ethanol solution.
4. Add water to bring the total volume up to 100 mL.

Agent B

Ethanol	40.0 mL
NaCl	0.9 g
Distilled or deionized water	≅60.0 mL

1. Dissolve 0.9 g NaCl in approximately 55 mL of distilled or deionized water.
2. Add the mixture to 40 mL of ethanol.
3. Add water to bring the total volume up to 100 mL.

GRAM STAIN REAGENTS

Gram Crystal Violet (Modified Hucker's)

Solution A

Crystal violet dye (90%)	2.0 g
Ethanol (95%)	20.0 mL

Solution B

Ammonium oxalate	0.8 g
Distilled or deionized water	80.0 mL

1. Combine solutions A and B. Store for 24 hours.
2. Filter before use.

Gram Decolorizer

Ethanol (95%)

Gram Iodine

Potassium iodide	2.0 g
Iodine crystals	1.0 g
Distilled or deionized water	300.0 mL

1. Dissolve the potassium iodide in the water *first*.
2. Dissolve the iodine crystals in the solution.
3. Store in an amber bottle.

Gram Safranin

Safranin O	0.25 g
Ethanol (95%)	10.0 mL
Distilled or deionized water	100.0 mL

1. Dissolve the safranin O in the ethanol.
2. Add the water.

NEGATIVE STAIN

Nigrosin

Nigrosin	10.0 g
Distilled or deionized water	100.0 mL

SIMPLE STAINS

Crystal Violet (See Gram Stain)

Methylene Blue (See Acid Fast, Hot)

Safranin (See Gram Stain)

Carbolfuchsin (See Acid Fast, Hot)

SPORE STAIN

Malachite Green

Malachite green dye	5.0 g
Distilled or deionized water	100.0 mL

Safranin (See Gram Stain)

VOGEL-BONNER SALTS (50X)

Magnesium sulfate	10.0 g
Citric acid	100.0 g
Dipotassium phosphate	500.0 g
Monosodium ammonium phosphate	175.0 g
Distilled or deionized water to bring volume to	1.0 liter*

* The solid ingredients raise the solvent level significantly in this preparation. Use only enough water initially to dissolve the ingredients then add water to bring the volume to one liter.

REFERENCES

Baron, Ellen Jo, Lance R. Peterson, and Sydney M. Finegold. 1994. *Bailey & Scott's Diagnostic Microbiology*, 9th Ed. Mosby–Yearbook, Inc., St. Louis, MO.

DIFCO Laboratories. 1984. *DIFCO Manual*, 10th Ed. DIFCO Laboratories, Detroit, MI.

Eisenstadt, Bruce C. Carlton, and Barbara J. Brown. 1994. *Methods for General and Molecular Bacteriology*, edited by Philipp Gerhardt, R. G. E. Murray, Willis A. Wood, and Noel R. Krieg, American Society for Microbiology, Washington, DC.

Koneman, Elmer W., *et al.* 1997. *Color Atlas and Textbook of Diagnostic Microbiology*, 5th Ed. Lippincott-Raven Publishers, Philadelphia, PA.

MacFaddin, Jean F. 1980. *Biochemical Tests for Identification of Medical Bacteria*, 2nd Ed. Williams & Wilkins, Baltimore, MD.

Power, David A. and Peggy J. McCuen. 1988. *Manual of BBL® Products and Laboratory Procedures*, 6th Ed. Becton Dickinson Microbiology Systems, Cockeysville, MD.

APPENDIX B

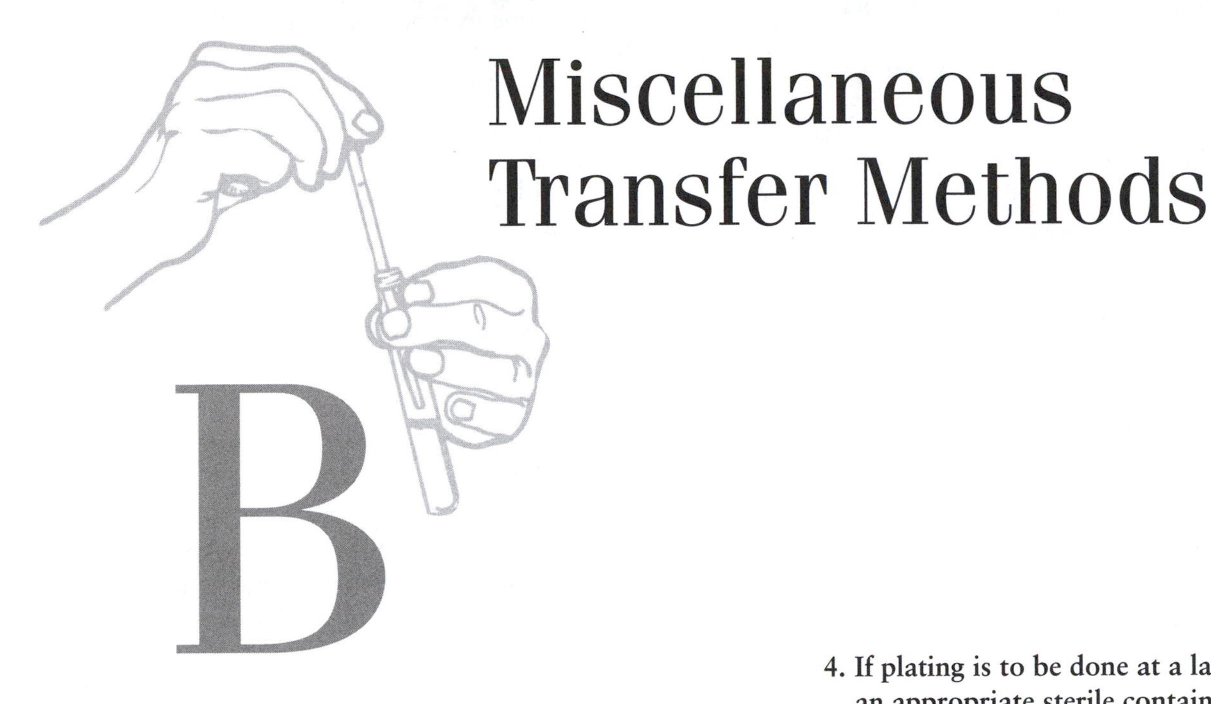

Miscellaneous Transfer Methods

Following are instructions for transfer methods that are less routinely performed than those in Exercise 2-1. As in Exercise 2-1, new skills in each process are printed in bold type.

Transfers Using a Sterile Cotton Swab

A sterile cotton swab is generally used to obtain a sample from a primary source—either a patient or an environmental site. Occasionally, swabs are used to transfer pure cultures. Sterile swabs may be dry or they may be in sterile water, depending on the sample source. In either case, care must be taken not to contaminate the swab by touching unintended surfaces with it. Your instructor may provide specific instructions on sample collection from sources other than the ones below.

Obtaining a Sample from a Patient's Throat with a Cotton Swab

1. **A sterile tongue depressor and swab prepared in sterile water may be used together to obtain a sample from the throat. Have the patient open his/her mouth, then gently press down on the tongue (Figure B-1).**
2. **With the swab in the dominant hand, carefully sample the patient's throat with a swirling motion. Touching other parts of the oral cavity is likely to cause contamination. Also, avoid touching the soft palate or a gag reflex may be initiated!**
3. **Transfer the sample to an appropriate plated medium (see Exercise 1-3) as quickly as possible.**
4. **If plating is to be done at a later time, place the swab in an appropriate sterile container (such as a sterile, capped test tube).**

Obtaining an Environmental Sample with a Cotton Swab

1. **A sterile cotton swab prepared in sterile water may be used to obtain a sample from an environmental source.**
2. **Rotate the swab to collect from the area to be sampled (Figure B-2).**
3. **Transfer the sample to an appropriate plated medium (see Exercise 1-3) as quickly as possible.**
4. **If plating is to be done at a later time, place the swab in an appropriate sterile container (such as a sterile, capped test tube).**

■ Figure B-1 Taking a Throat Sample
Use a sterile tongue depressor and cotton swab to obtain a sample from a patient's throat. Be careful not to touch other parts of the oral cavity or the sample will get contaminated. Transfer the sample to a sterile medium as soon as possible.

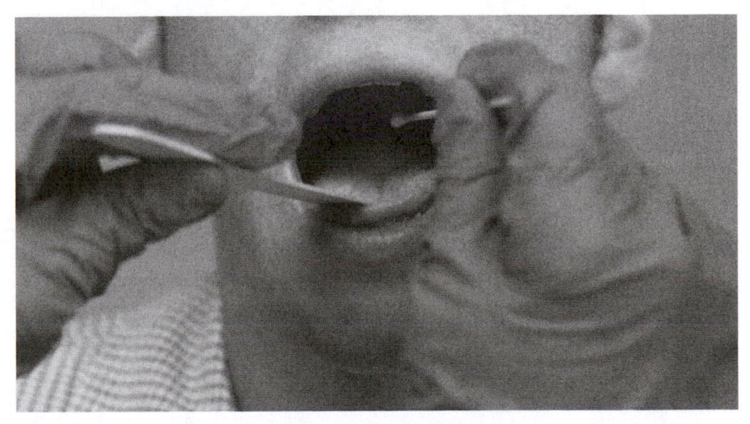

■ **Figure B-2 Taking an Environmental Sample**
Use a spinning motion of a sterile swab to sample inanimate objects in the environment. The swab may be placed in a sterile test tube until it is convenient to transfer the sample to a growth medium.

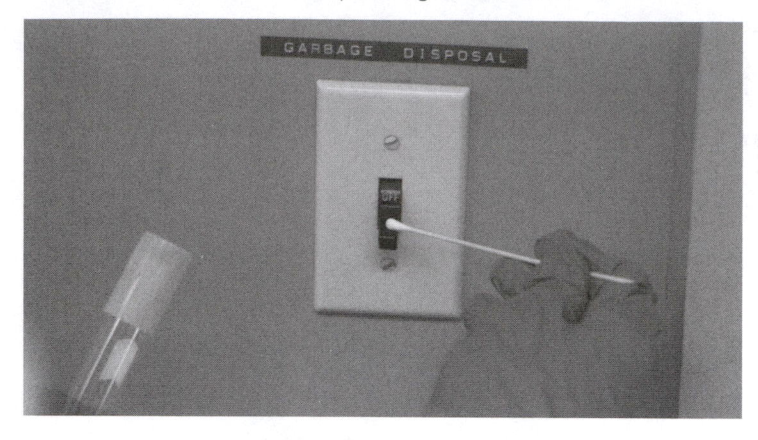

Stab Inoculation of Agar Tubes Using an Inoculating Needle

Stab inoculations of agar tubes are used for several types of differential media (usually to examine growth under anaerobic conditions or to observe motility). A stab is *not* used to produce a culture of microbes for transfer to another medium.

1. Remove the cap of the sterile medium with the little finger of your inoculating needle hand and hold it there.
2. Flame the tube by quickly passing it through the Bunsen burner flame a couple of times. Keep your needle hand still.
3. Hold the open tube on an angle to minimize airborne contamination. Keep your needle hand still.
4. **Carefully move the agar tube over the needle wire (Figure B-3). Insert the needle into the agar to about 1 cm from the bottom.**
5. **Withdraw the tube carefully so the needle leaves along the same path it entered. (Be especially careful when removing the tube not to catch the needle tip on the tube lip. This springing action of the needle creates bacterial aerosols.)**
6. Flame the tube lip as before. Keep your needle hand still.
7. Keeping the needle hand still (remember, it has growth on it), move the tube to replace its cap.
8. Sterilize the needle as before by incinerating it in the Bunsen burner flame. It is especially important to flame it from base to tip now because the needle has lots of bacteria on it.
9. Label the tube with your name, date, and organism. Incubate at the appropriate temperature for the assigned time.

■ **Figure B-3 Agar Deep Stab**
Use the inoculating needle to stab the agar to a depth about 1 cm from the bottom. It is generally desirable to remove the needle along the original stab line and not create a new one. Sterilize the needle upon completion.

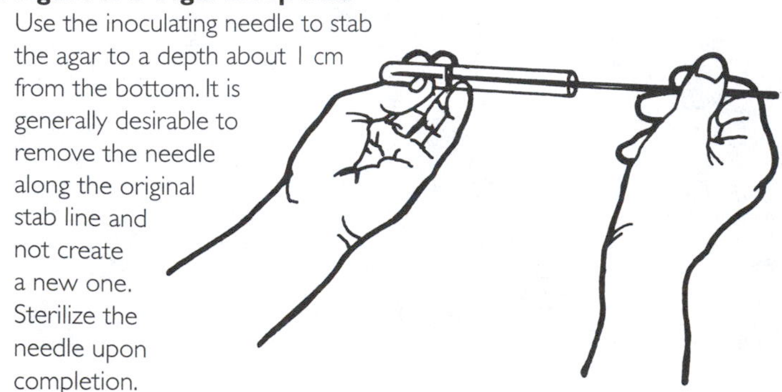

Spot Inoculation of an Agar Plate

Sometimes, an agar plate may be used to grow several different specimens at once. This is a typical practice with plated *differential* media (*i.e.*, media designed to differentiate organisms based on growth characteristics). Prior to beginning the transfer, the plate may be divided into as many as four sectors using a marking pen. (Some plates already have marks on the base for this purpose.) Each may then be inoculated with a different organism. Inoculation involves touching the loop to the agar surface once so that growth is restricted to a single spot—hence the name "spot inoculation."

1. Lift the lid of the sterile agar plate and use it as a shield to prevent airborne contamination.
2. **Touch the agar surface towards the periphery of the sector (Figure B-4).**
3. Remove the loop and replace the lid.
4. Sterilize your loop as before. It is especially important to flame it from base to tip now because the loop has lots of bacteria on it.
5. Label the plate's base with your name, date, and organism(s) inoculated.
6. Incubate the plate in an inverted position for the assigned time at the appropriate temperature.

■ **Figure B-4 Spot Inoculation of a Plate**
Each of four sectors is spot inoculated with a different organism by touching the loop to the surface and making a mark about 1 cm in length. Generally, spot inoculations are done toward the edge (rather than the crowded middle) to prevent overlapping growth and/or test results.

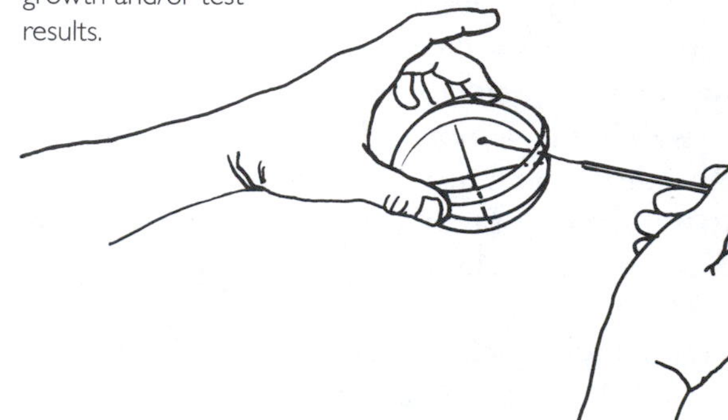

APPENDIX C

Use of Glass Pipettes

Transfers from a Broth Culture Using a Glass Pipette

Glass pipettes are used to transfer a known volume of liquid diluent, media, or culture. Originally, pipettes were filled by sucking on them like a drinking straw, but mouth pipetting is dangerous and has been replaced by mechanical pipettors. Three examples are shown in Figure C-1, each with its own method of operation. Your instructor will show you how to properly use the style of pipettor available in your lab.

In order to use a pipette correctly, you must be able to correctly read the calibration. Examine Figure C-2. The numbers indicate the pipette's *total volume* and its *smallest calibrated increments*. This is a 5.0 mL pipette divided into 0.1 mL increments.

When reading volumes, use the base of the meniscus (Figure C-3). The volume in the center pipette is read at exactly 3.0 mL because the meniscus is resting on the line.

■ Figure C-1 Mechanical Pipettors
Three examples of mechanical pipettors are shown here, each with its own method of operation. Your instructor will show you how to properly use the style of pipettor available in your lab. From top to bottom: A pipette filler/dispenser, a pipette bulb, and a plastic pump.

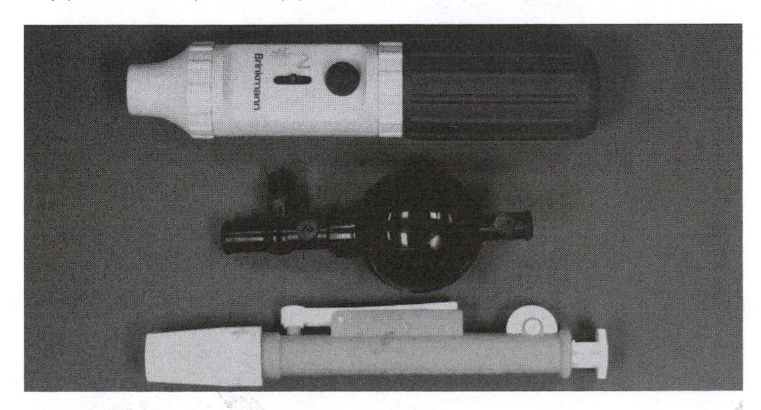

■ Figure C-2 Pipette Calibration
Read the pipette calibration prior to using a pipette. The numbers indicate the pipette's *total volume* and its *smallest calibrated increments*. This is a 5.0 mL pipette divided into 0.1 mL increments.

■ Figure C-3 Read the Base of the Meniscus
When reading volumes, use the base of the meniscus. The volume in the center pipette is read at exactly 3.0 mL because the meniscus is resting on the line. The left pipette is read as 2.9 mL and the right pipette is read as 3.1 mL (0 is always at the pipette's top).

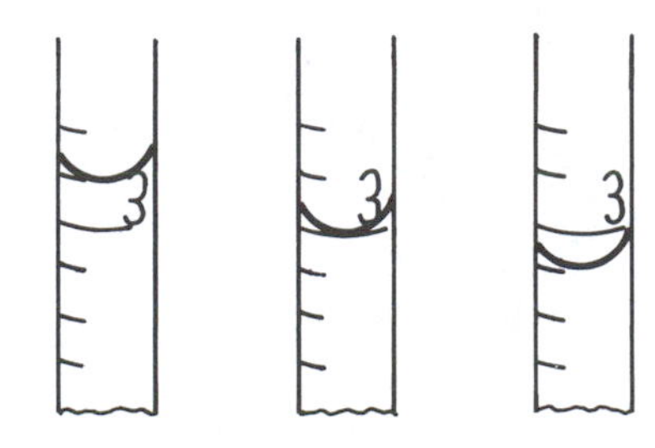

The left pipette is read as 2.9 mL and the right pipette is read as 3.1 mL (0 is always at the pipette's top). Although the difference in volume between these three pipettes may seem negligible, it may be enough to introduce significant error into your work.

Two pipette styles are used in microbiology (Figure C-4). These are the **serological pipette** and the **Mohr pipette.** A serological pipette is calibrated *to deliver* (TD) its volume by completely draining it and blowing out the last drop. The tip of a Mohr pipette is not graduated, so fluid flow must be stopped at a calibration line. Stopping the fluid beyond the last line on a Mohr pipette results in an unknown volume being dispensed. In either case, volumes are read at the bottom of the meniscus of fluid.

Important: If pipetting a bacterial culture, be careful not to allow any to drop from the pipette before disposing of it in the autoclave container. Clean up any spills.

Filling a Glass Pipette

1. Bacteria should be suspended in the broth with a vortex mixer (Figure 1-6) or by agitating with your fingers (Figure 1-7). Be careful not to splash the broth into the cap or lose control of the tube.
2. **Pipettes are sterilized in metal canisters or packages (if disposable) and are stored in groups of a single size (Figure C-5).** ***Be sure you know what volume your pipette will deliver.*** **Set the canister at the table edge and remove its lid. (Canisters should not be stored in an upright position as they may fall over and break the pipettes or become contaminated.) If using pipettes in a package, open the end** ***opposite the tips.*** **Grasp** ***one pipette only*** **and remove it.**
3. **Carefully insert the pipette into the mechanical pipettor (Figure C-6). It's best to grasp the pipette near the end with your fingertips. This gives you more control and reduces the chance that you will break the pipette and cut your hand.** ***Do not touch*** **any part of the pipette that will contact the specimen or the medium or you risk introducing a contaminant. Also, do not lay the pipette on the tabletop while you continue.**
4. While keeping the pipette hand still, bring the culture tube towards it. Use your little finger to remove and hold its cap.
5. Flame the open end of the tube by passing it through a Bunsen burner flame two or three times.
6. Hold the tube at an angle to prevent contamination from above.
7. **Insert the pipette and withdraw the appropriate volume (Figure C-7). Bring the pipette to a vertical position briefly to accurately read the meniscus. (Remember: the volumes in the pipette are correct only if the meniscus of the fluid inside is resting** ***on*** **the line, not below it.) Then carefully remove the pipette from the tube.**

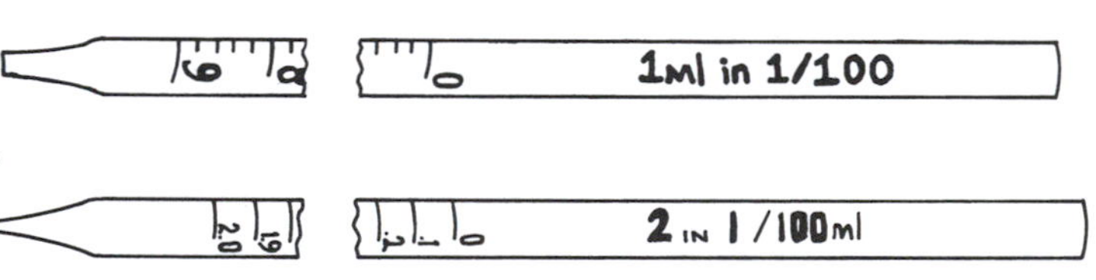

■ Figure C-4 Two Types of Pipettes
Two pipette styles, the *serological pipette* (top) and the Mohr pipette (bottom), are used in microbiology. A serological pipette is calibrated *to deliver* (TD) its volume by completely draining it and blowing out the last drop. The tip of a Mohr pipette is not graduated, so fluid flow must be stopped at a calibration line. Stopping the fluid beyond the last line on a Mohr pipette results in an unknown volume being dispensed.

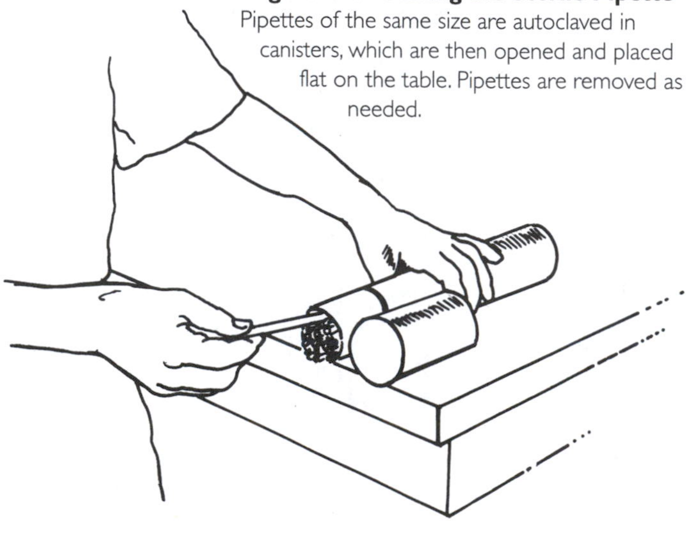

■ Figure C-5 Getting the Sterile Pipette
Pipettes of the same size are autoclaved in canisters, which are then opened and placed flat on the table. Pipettes are removed as needed.

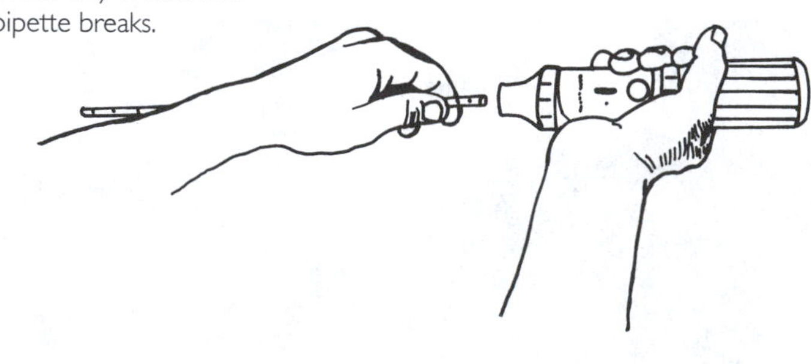

■ Figure C-6 Assembling the Pipette
Carefully insert the pipette into a mechanical pipettor. Notice that the pipette is held near the end with the fingertips. For safety, the hand is out of the way in case the pipette breaks.

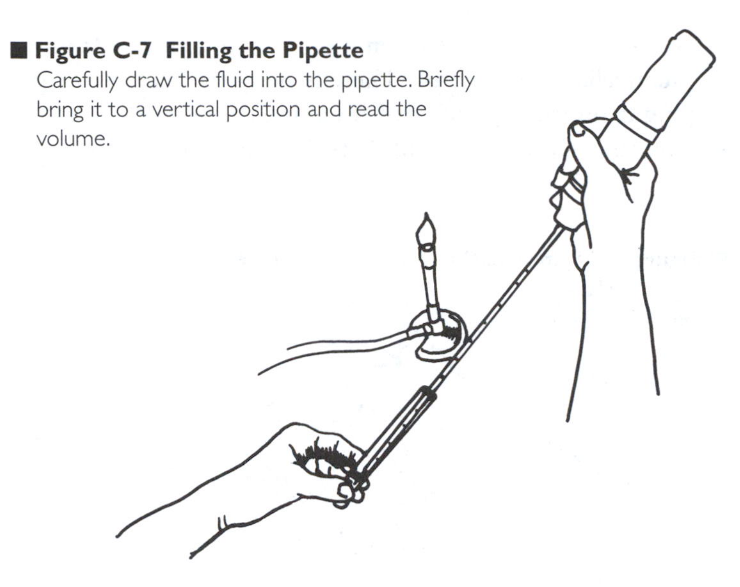

■ **Figure C-7 Filling the Pipette**
Carefully draw the fluid into the pipette. Briefly bring it to a vertical position and read the volume.

8. Flame the tube lip as before. Keep your pipette hand still.
9. Keeping the pipette hand still (remember, it contains fluid with microbes in it), move the tube to replace its cap.
10. What you do next depends on the medium to which you are transferring the growth. Please continue with the appropriate inoculation section.

Inoculation of Broth Tubes with a Pipette

Pipettes are often used to inoculate a known volume of culture into a known volume of diluent during serial dilutions.

1. While keeping the pipette hand still, bring the broth tube towards it. Use your little finger to remove and hold its cap.
2. Flame the tube by quickly passing it through the Bunsen burner flame two or three times. Keep your pipette hand still.
3. Hold the open tube on an angle to minimize airborne contamination. Keep your pipette hand still.
4. **Insert the pipette tip and dispense the correct volume of inoculum (Figure C-8).**
5. **Withdraw the tube from over the pipette. Before completely removing it, touch the pipette tip to the glass to remove any excess broth (Figure C-9).**

■ **Figure C-8 Inoculate the Broth**
Hold the open tube at an angle to prevent aerial contamination. Insert the pipette tip and dispense the correct volume of inoculum.

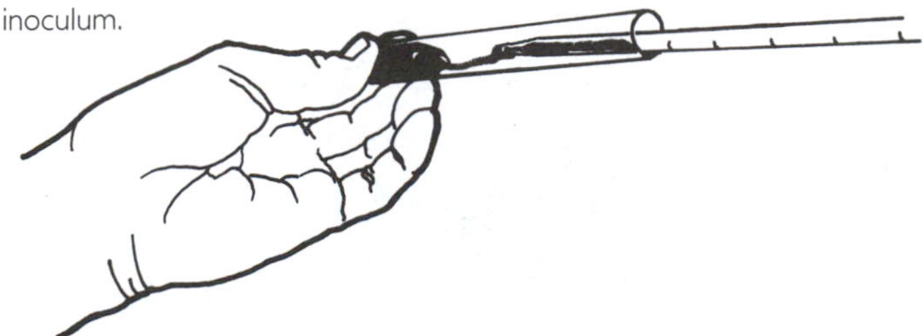

6. **Completely remove the pipette, but avoid waving it around. This can create aerosols.**
7. **Flame the tube lip as before. Keep your pipette hand still.**
8. Keeping the pipette hand still, move the tube to replace its cap.
9. **The pipette is contaminated with microbes and *must be* correctly disposed of. Each lab has its own specific procedures and your instructor will advise you what to do. Glass pipettes are typically placed in pipette disposal container containing a small amount of disinfectant until they are autoclaved and reused (Figure C-10). Disposable pipettes must be placed in an appropriate biohazard container. In either case, be careful when removing the pipette from the mechanical pipettor. There is danger of culture dripping from the pipette or of breaking the glass.**

■ **Figure C-9 Remove Excess Broth**
Withdraw the tube from over the pipette. Before completely removing it, touch the pipette tip to the glass to remove any excess broth. Completely remove the pipette, but avoid waving it around. This can create aerosols.

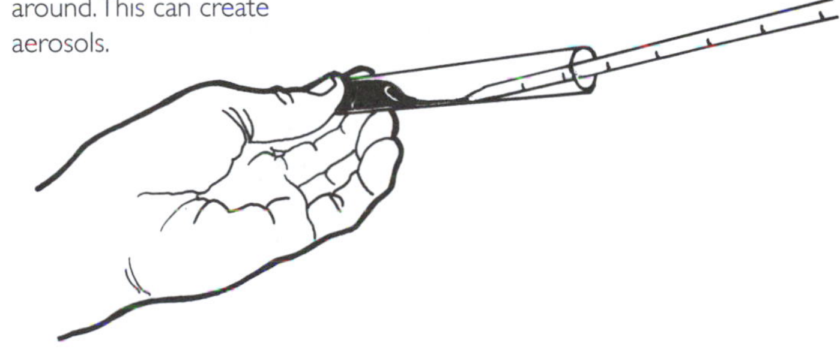

■ **Figure C-10 Dispose of the Pipette**
The pipette is contaminated with microbes and must be correctly disposed of. Each lab has its own specific procedures and your instructor will advise you what to do. Shown here is a glass pipette being placed in pipette disposal container containing a small amount of disinfectant. Disposable pipettes must be placed in an appropriate biohazard container. In either case, be careful when removing the pipette from the mechanical pipettor. There is danger of culture dripping from the pipette or of breaking the glass.

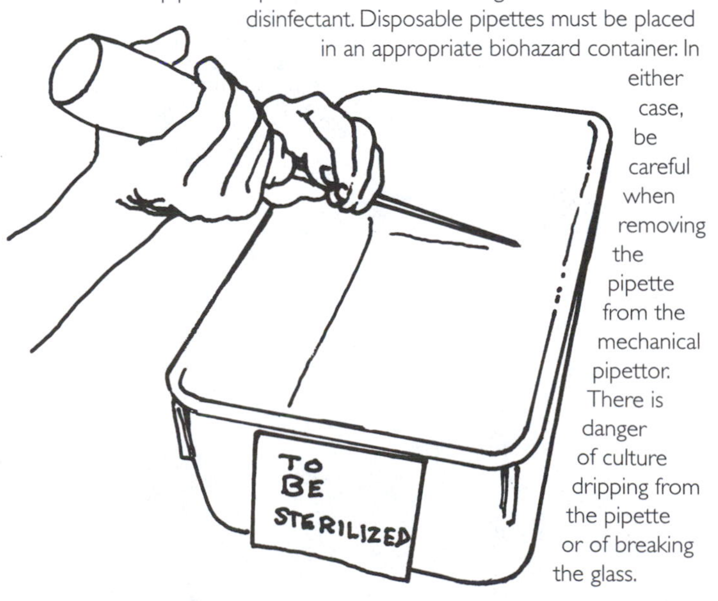

Inoculation of Agar Plates with a Pipette

1. Lift the plate's lid and use it as a shield to protect from airborne contamination.
2. **Hold the pipette over the agar and dispense the correct volume (often 0.1 mL) onto the center of the agar surface (Figure C-11). From this point, the remainder of steps should be completed within about 15 seconds to prevent the inoculum from soaking into the agar.**
3. **The pipette is contaminated with microbes and must be correctly disposed of. Each lab has its own specific procedures and your instructor will advise you what to do. Glass pipettes are typically placed in pipette disposal container containing a small amount of disinfectant until they are autoclaved and reused. Disposable pipettes must be placed in an appropriate biohazard container. In either case, be careful when removing the pipette from the mechanical pipettor. There is danger of culture dripping from the pipette or of breaking the glass.**
4. Continue with the Spread Plate Technique (Exercise 1-4).

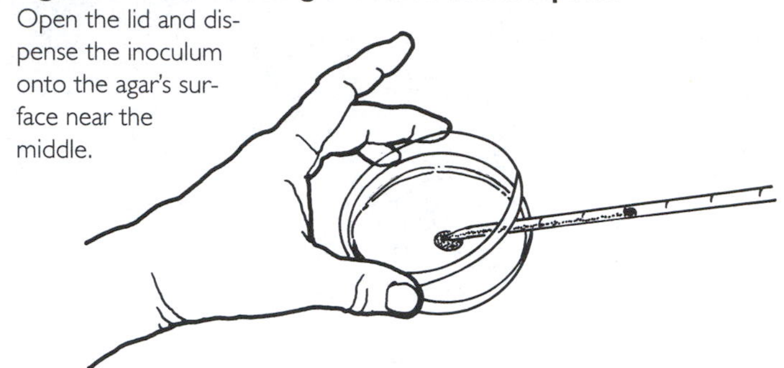

■ Figure C-11 Inoculating the Plate with a Pipette
Open the lid and dispense the inoculum onto the agar's surface near the middle.

APPENDIX D

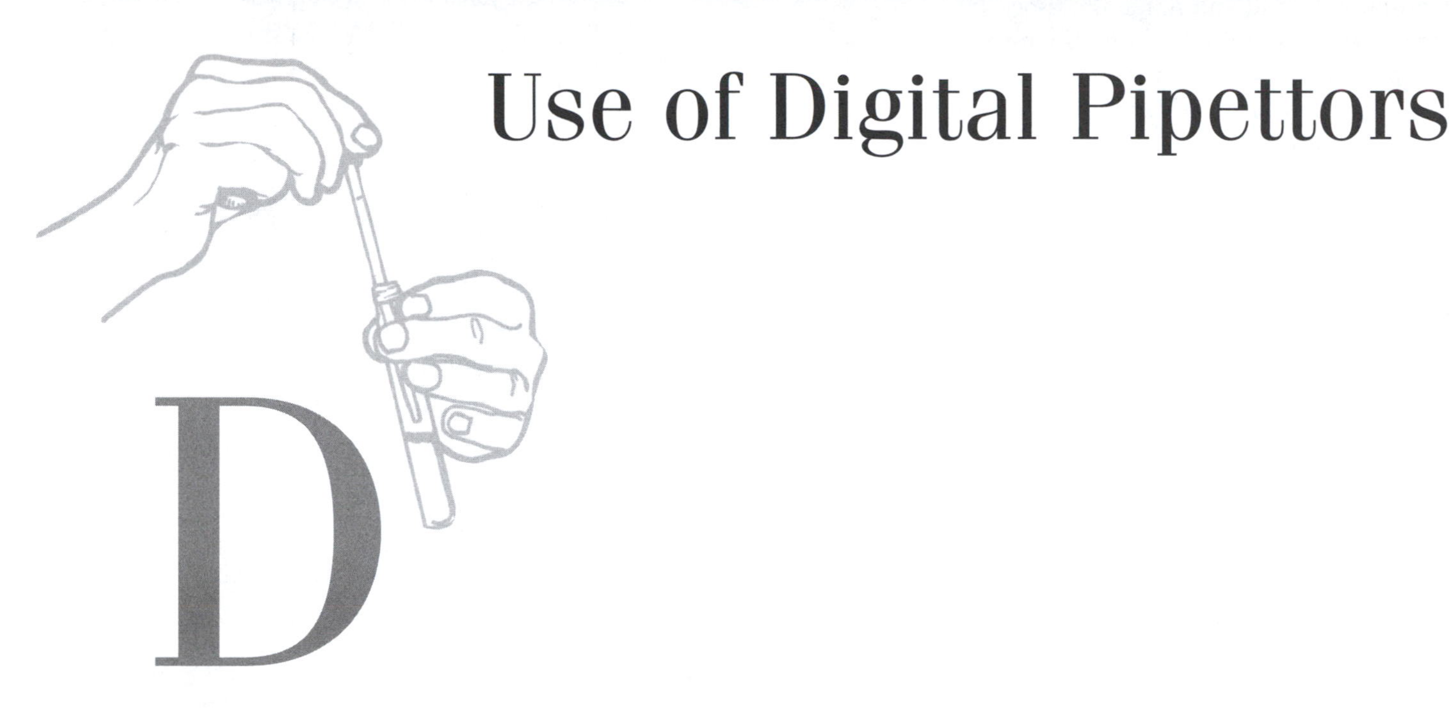

Use of Digital Pipettors

Transfers from a Broth Culture Using a Digital Pipette

Modern molecular biology procedures often involve transferring extremely small volumes of liquid with great precision and accuracy. This has led to the development of digital pipettors (Figure D-1) that can be set to dispense microliter volumes up through milliliter volumes. Common digital pipettors are calibrated to dispense volumes of 1 to 100 µL, 100 to 1000 µL, or 1 to 5 mL or more.

Filling a Digital Pipettor

Many manufacturers make digital pipettors, but they all work basically the same way. The following instructions are for Eppendorf Series 2100 models.

1. Growth may be suspended in the broth with a vortex mixer (Figure 1-6) or by agitating with your fingers (Figure 1-7). Be careful not to splash the broth into the cap or lose control of the tube.
2. **Determine which digital pipettor should be used to dispense the desired volume.**
3. **Turn the setting ring to set the desired volume (Figure D-2). Never turn the setting ring past the maximum volume for the pipettor or you may damage it.**
4. **Hold the digital pipettor in your dominant hand.**
5. **Open the rack of appropriate pipette tips for your pipettor (these are color-coded and match the pipettor) and push the pipettor into a sterile tip (Figure D-3). Close the rack. Never touch the pipette tip with your hands or leave the rack open.** ***Never use a digital pipettor without a tip.***

■ Figure D-1 Digital Pipettor
Shown is a digital pipettor to be used for dispensing volumes between 100 and 1000 µL. Always use a digital pipettor with the appropriate tip.

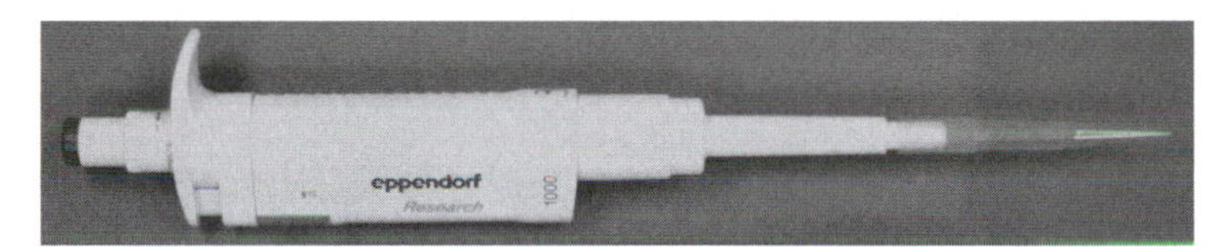

■ Figure D-2 Setting the Volume
Rotating the setting ring sets the volume on a digital pipettor. This pipettor has been set at 90.0 µL (see inset—the horizontal line between the third and fourth numerals is a decimal point). Never turn the setting ring beyond the maximum volume for the pipette or you may damage it.

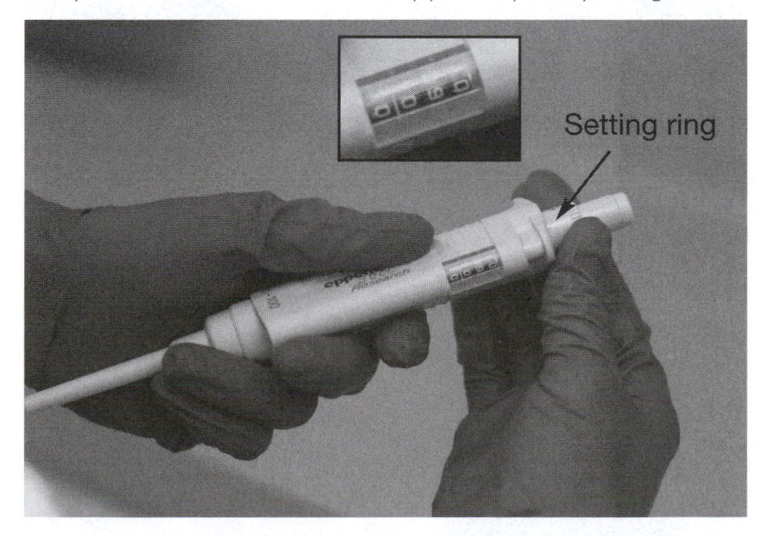

6. Remove the culture tube's cap with the little finger of your dominant hand and flame the tube.
7. **Press the control button down with your thumb to the first stop—the measuring stroke.**
8. **Insert the tip into the broth approximately 3 mm while holding it vertically (Figure D-4).**
9. **Slowly release pressure with your thumb to draw fluid into the tip. Be careful not to pull any air into the tip.**
10. Flame the tube as before.
11. Keeping the pipettor hand still, move the tube to replace its cap.
12. What you do next depends on the medium to which you are transferring the growth. Please continue with the appropriate inoculation section.

Inoculation of Broth Tubes With a Digital Pipettor

1. Remove the cap of the sterile medium with the little finger of your pipettor hand and hold it there.
2. Flame the tube by quickly passing it through the Bunsen burner flame a couple of times. Keep your pipettor hand still.
3. **Insert the pipette tip into the tube. Hold it at an angle against the inside of the glass (Figure D-5).**

■ Figure D-3 Digital Pipettor Tips
Digital pipettors must be fitted with a sterile tip of appropriate size (color-coded to match the pipettor). Open the case and press the pipettor into a tip, then close the case to maintain sterility. Do not touch the pipettor tip.

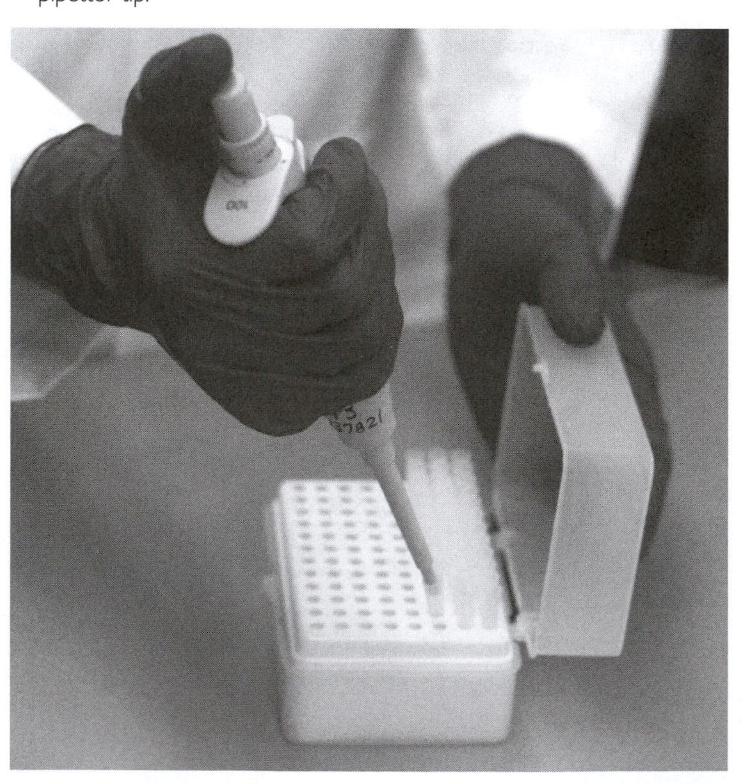

■ Figure D-4 Filling the Pipette
Depress the control button with your thumb to the first stop (measuring stroke). Holding the tube and pipettor in a vertical position, insert the tip into the fluid to a depth of 3 mm (see inset). Slowly release pressure with your thumb to fill the pipettor.

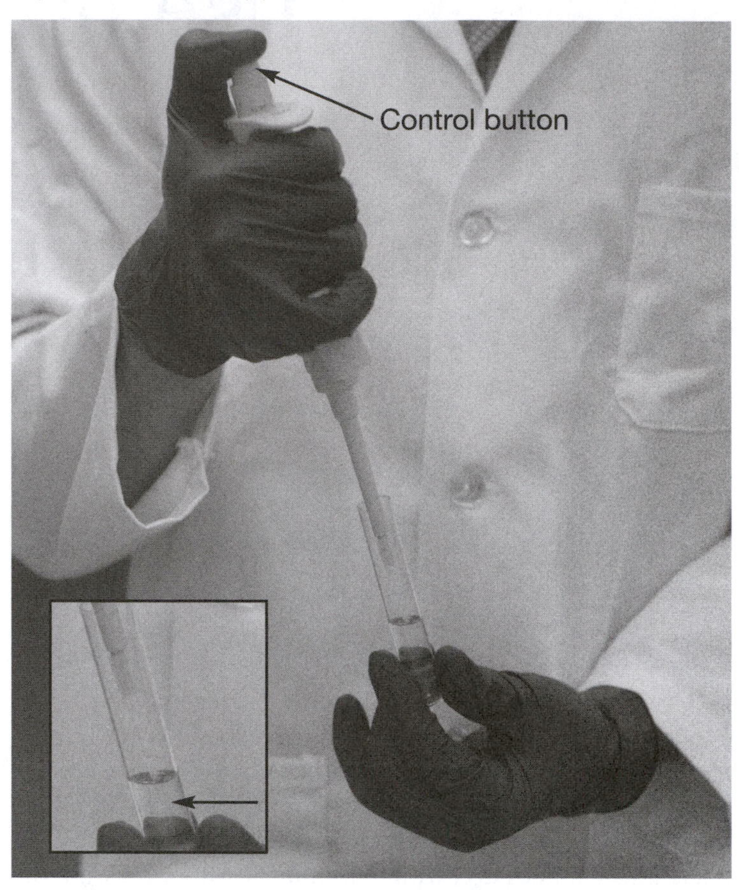

■ Figure D-5 Dispensing to a Liquid
Hold the pipettor on an angle with the tip against the glass. Press the control button to the first stop. When no more fluid comes out, continue pressing to the second stop. Slowly release pressure on the control button, then eject the tip into a biohazard container.

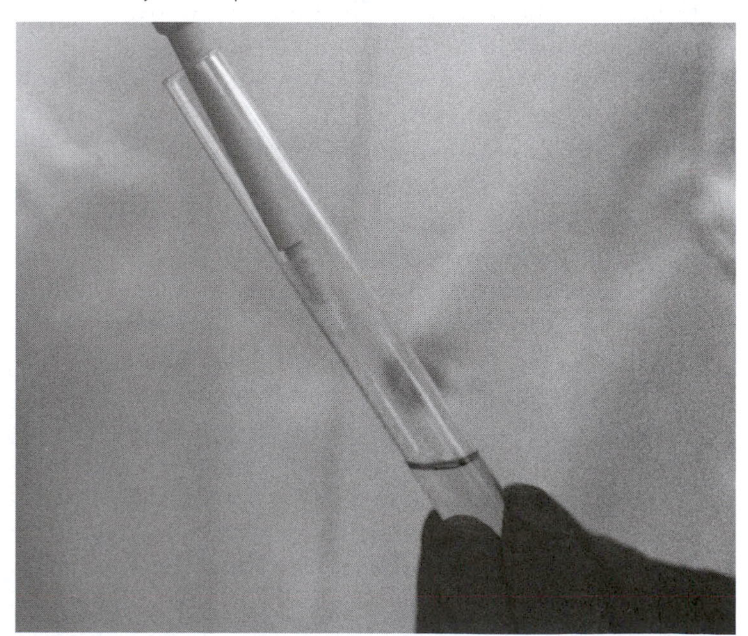

4. **Depress the control button slowly with your thumb to the first stop and pause until no more liquid is dispensed. Then, continue pressing to the second stop (blow-out) to deliver the remaining volume.**
5. **While keeping pressure on the control button, carefully remove the pipettor from the tube by sliding it along the glass. Once it is out of the tube, slowly release pressure on the control button.**
6. Flame the tube lip as before. Keep your pipette hand still.
7. Keeping the pipette hand still, move the tube to replace its cap.
8. **The pipettor tip is contaminated with microbes and must be correctly disposed of. Use the ejector button to remove the tip into an appropriate biohazard container (Figure D-6). Each lab has its own specific procedures and your instructor will advise you what to do.**

■ **Figure D-6 Ejecting the Tip**
The contaminated tip should be ejected into an appropriate biohazard container using the tip ejector button.

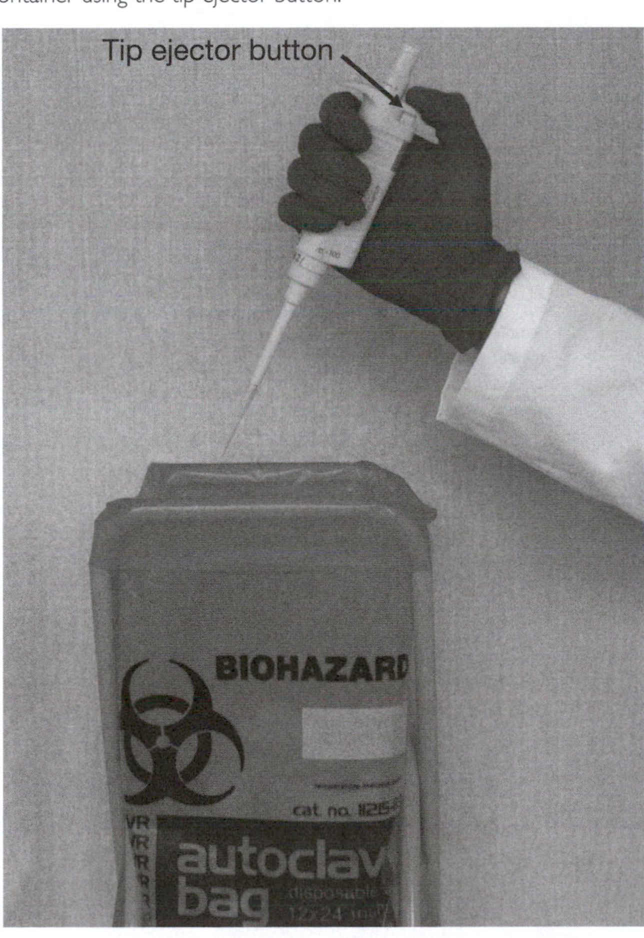

Inoculation of Agar Plates with a Pipettor

1. Lift the plate's lid and use it as a shield to protect from airborne contamination.
2. **Place the pipette tip over the agar surface. Be sure to hold the pipettor in a vertical position. (Figure D-7).**
3. **Depress the control button slowly with your thumb to the first stop and pause until no more liquid is dispensed. Then, continue pressing to the second stop (blow-out) to deliver the appropriate volume. From this point, the remainder of steps should be completed within about 15 seconds to prevent the inoculum from soaking into the agar.**
4. **The pipettor tip is contaminated with microbes and must be correctly disposed of. Use the ejector button to remove the tip into an appropriate biohazard container (Figure D-6). Each lab has its own specific procedures and your instructor will advise you what to do.**
5. **Continue with the Spread Plate Technique (Exercise 1-4).**

■ **Figure D-7 Dispensing to a Plate**
Hold the pipette in a vertical position and press the control button to the first stop. When no more fluid comes out, continue pressing to the second stop. Slowly release pressure on the control button, then eject the tip into a biohazard container.

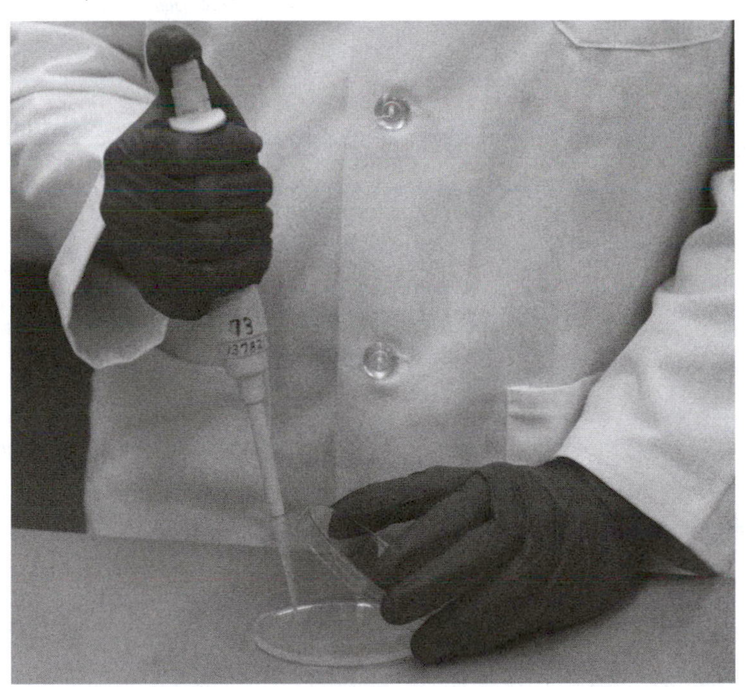

REFERENCES

Brinkman Instruments, Inc. *Eppendorf Series 2001 Pipette Instruction Manual.* One Cantiague Road, Westbury, New York 11590-0207.

APPENDIX E

Morphological Unknown Flowchart

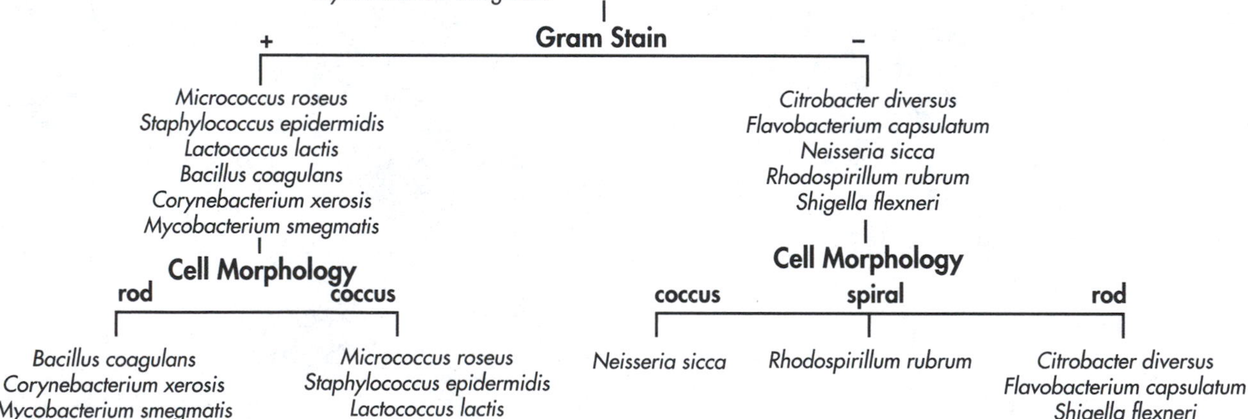